Culture and Sustainability in European Cities

European cities are contributing to the development of a more sustainable urban system that is capable of coping with economic crises, ecological challenges, and social disparities in different nation-states and regions throughout Europe.

This book reveals in a pluralistic way how European cities are generating new approaches to their sustainable development, and the special contribution of culture to these processes. It addresses both a deficit of attention to small and medium-sized cities in the framework of European sustainable development and an underestimation of the role of culture, artistic expression, and creativity for integrated development of the city as a prerequisite to urban sustainability. On the basis of a broad collection of case studies throughout Europe, representing a variety of regionally specific cultural models of sustainable development, the book investigates how participative culture, community arts, and, more generally, creativity of civic imagination are conducive to the goal of a sustainable future for small and medium-sized cities.

This is an essential volume for researchers and postgraduate students in urban studies, cultural studies, cultural geography, and urban sociology, as well as for policy makers and practitioners wanting to understand the specificity of European cities as hubs of innovation, creativity, and artistic industriousness.

Svetlana Hristova is Associate Professor in Sociology of Culture and Cultural Anthropology at the Faculty of Arts in the South-West University in Blagoevgrad, Bulgaria, and Chair of the ENCATC Thematic Area "Urban Management and Cultural Policies of City."

Milena Dragićević Šešić is Head of the UNESCO Chair in Interculturalism, Art Management and Mediation, and Professor of Cultural Policy and Cultural Management, Cultural Studies, and Media Studies at the University of Arts, Belgrade, Serbia.

Nancy Duxbury is Senior Researcher at the Centre for Social Studies, University of Coimbra, Portugal, and Co-Coordinator of its Cities, Cultures and Architecture Research Group. She is also Adjunct Professor of the School of Communication, Simon Fraser University, Canada.

Routledge Studies in Culture and Sustainable Development

Culture as an aspect of sustainability is a relatively new phenomenon but is beginning to attract attention among scholars and policy makers. This series opens up a forum for debate about the role of culture in sustainable development, treating culture and sustainability as a meta-narrative that will bring together diverse disciplines. Key questions explored in this series will include: how should culture be applied in sustainability policies; what should be sustained in culture; what should culture sustain; and what is the relationship of culture to other dimensions of sustainability?

Books in the series will have a variety of geographical foci and reflect different disciplinary approaches (for example, geography, sociology sustainability science, environmental and political sciences, anthropology, history, archaeology, and planning). The series will be addressed in particular to postgraduate students and researchers from a wide cross-section of disciplines.

Series Editors:
Katriina Soini, University of Jyväskylä, Finland, and
Natural Resources Institute Finland
Joost Dessein, ILVO, Ghent University, Belgium

Culture and Sustainability in European Cities
Imagining Europolis
Edited by Svetlana Hristova, Milena Dragićević Šešić, and Nancy Duxbury

Theory and Practice in Heritage and Sustainability
Between past and future
Edited by Elizabeth Auclair and Graham Fairclough

Cultural Sustainability and Regional Development
Theories and practices of territorialisation
Edited by Joost Dessein, Elena Battaglini, and Lummina Horlings

"'Imagining Europolis' comes at a moment when more than ever we need – at EU level – to collectively build a common civic space where European citizens can discuss and reflect on crucial issues for the further development of our European project. The book clearly contributes to this endeavour. My thanks to the authors!"

– Sylvain Pasqua, Senior Expert in EU Cultural Policies in the European Commission, Coordinator of the "European Capitals of Culture" Scheme of the European Union

"This is a very timely book that gathers many distinguished contributors on a hot topic such as sustainable culture-led development in small and medium-sized European cities – the true backbone of Europe. Research, fresh ideas, and experiences in this field are much needed in the Europe of today. Warmly recommended reading."

– Pier Luigi Sacco, Professor of Cultural Economics, IULM University, Milan, Italy; Member of the Technical-Scientific Committee on Cultural Economics of the Italian Ministry of Culture and Heritage

"The complex cultural dimensions of the urban condition today and of the theory and practice of sustainability are topics hampered by commonplace as well as wishful if not incantatory thinking. The essays in this truly significant volume on European cities break exciting new ground, however. They provide a rich template for far more informed and fruitful debate."

– Yudhishthir Raj Isar, Professor of Cultural Policy Studies, The American University of Paris, France; Institute for Culture and Society, University of Western Sydney, Australia

"'Imagining Europolis' shows compellingly how a number of the continent's small and medium-sized cities offer key creative, imaginative, and participatory models for a culture-based sustainable urbanism while clearly theorizing ever-present tensions and faultlines. In its models for sustainable city co-creation the book offers valuable lessons for cities everywhere, including the urban behemoth Cosmopolis."

– Greg Young, Associate Professor of Urban and Regional Planning, University of Sydney, Australia

Culture and Sustainability in European Cities

Imagining Europolis

Edited by Svetlana Hristova,
Milena Dragićević Šešić, and
Nancy Duxbury

Routledge
Taylor & Francis Group
LONDON AND NEW YORK

earthscan
from Routledge

First published 2015
by Routledge

2 Park Square, Milton Park, Abingdon, Oxfordshire OX14 4RN
52 Vanderbilt Avenue, New York, NY 10017

Routledge is an imprint of the Taylor & Francis Group, an informa business

First issued in paperback 2020

British Library Cataloguing-in-Publication Data
A catalogue record for this book is available from the British Library

Library of Congress Cataloging-in-Publication Data
Cultural sustainability in European cities : imagining Europolis / edited by
Svetlana Hristova, Milena Dragićević Šešić, and Nancy Duxbury.
pages cm
Includes bibliographical references and index.
1. Europe—Cultural policy. 2. Europe—Civilization. 3. Urban policy—Europe. 4. Sustainability—Europe. 5. Sustainable urban development—Europe. I. Hristova Svetlana. II. Dragićević Šešić, Milena, 1954– III. Duxbury, Nancy.
D1055.C826 2015
306.094—dc23
2015001186

ISBN: 978-1-138-77841-2 (hbk)
ISBN: 978-0-367-66888-4 (pbk)

Typeset in Goudy
by Apex CoVantage, LLC

Contents

COST – European Cooperation in Science and Technology

This publication is supported by COST.

COST – European Cooperation in Science and Technology is an intergovernmental framework aimed at facilitating the collaboration and networking of scientists and researchers at the European level. It was established in 1971 by 19 member countries and currently includes 35 member countries across Europe, and Israel as a cooperating state.

COST funds pan-European, bottom-up networks of scientists and researchers across all science and technology fields. These networks, called 'COST Actions', promote international coordination of nationally-funded research. By fostering the networking of researchers at an international level, COST enables breakthrough scientific developments leading to new concepts and products, thereby contributing to strengthening Europe's research and innovation capacities.

COST's mission focuses in particular on:

- *Building capacity by connecting high quality scientific communities throughout Europe and worldwide;*
- *Providing networking opportunities for early career investigators;*
- *Increasing the impact of research on policy makers, regulatory bodies and national decision makers as well as the private sector.*

Through its inclusiveness policy, COST supports the integration of research communities in less research-intensive countries across Europe, leverages national research investments and addresses societal issues.

Over 45,000 European scientists benefit from their involvement in COST Actions on a yearly basis. This allows the pooling of national research funding and helps countries' research communities achieve common goals.

As a precursor of advanced multidisciplinary research, COST anticipates and complements the activities of EU Framework Programmes, constituting a "bridge" towards the scientific communities of emerging countries.

Traditionally, COST draws its budget for networking activities from successive EU RTD Framework Programmes.

EUROPEAN COOPERATION
IN SCIENCE AND TECHNOLOGY

COST is supported by the EU Framework Programme Horizon 2020.

Series introduction

Achieving a more sustainable level of development is the biggest global challenge of the twenty-first century, and new approaches are urgently needed to ensure that development is much better aligned with the environmental, societal, and economic problems we are facing. Scholars and policy makers have recognized the contribution of culture in sustainable development. The issue of culture in sustainable development is also being increasingly discussed in debates in various international, national, and local arenas, and there are ample local actor-driven initiatives. Yet despite this growing attention, there have been only a very few attempts to consider culture in a more analytical and explicit way within the frames of sustainability. The challenge of incorporating culture in sustainable development discourse, both scientifically and politically, is derived from the complex, normative, and multidisciplinary character of both culture and sustainable development. This difficulty should not, however, be any excuse for ignoring the cultural dimension within sustainable development.

The series "Routledge Studies in Culture and Sustainable Development" aims to analyse the diverse and multiple roles that culture plays in sustainable development. It takes as one of its starting points the idea that culture in sustainability serves as a "meta-narrative" which will bring together ideas and standpoints from an extensive body of academic research currently scattered among different disciplines and thematic fields. Moreover, the series responds to the strengthening call for inter- and transdisciplinary approaches, which is being heard in many quarters, but in few fields more strongly than that of sustainability and sustainable development, with its complex and systemic problems. By combining and confronting the various approaches – in both the sciences and the humanities and in dealing with social, cultural, environmental, political, and aesthetic disciplines – the series offers a comprehensive contribution to the present-day sustainability sciences as well as related policies.

The books in the series will take a broad approach to culture, giving space to all the possible understandings of culture from art-based definitions to way-of-life based approaches, and beyond. Furthermore, culture is not seen only as an additional aspect of sustainable development – as a "fourth pillar" – but, rather, as a mediator, a cross-cutting transversal framework, or even as a new set of guiding principles for sustainable development research, policies, and practices.

The essence of culture in, for, and as sustainable development will be explored through the series in various thematic contexts, representing a wide range of practices and processes (e.g. everyday life, livelihoods and lifestyles, landscape, artistic practices, aesthetic experiences, heritage, tourism). These contexts may concern urban, peri-urban, or rural contexts, and regions with different trajectories of socioeconomic development. The perspectives of the books will stretch from local to global and cover different temporal scales from past to present and future. These issues are valorized by theoretical or empirical analysis; their relationship to the ecological, social, and economic dimensions of sustainability will be explored, when appropriate.

The idea for the series is derived from the European COST Action IS1007 "Investigating Cultural Sustainability," running between 2011 and 2015. This network is comprised of a group of around one hundred researchers from twenty-six European countries, and representing many different disciplines. They have brought together their expertise, knowledge, and experience, and based on that they have built up new inter- and transdisciplinary understanding and approaches that can enhance and enrich research into culture in sustainable development, and support the work of the policy makers and practitioners in this field.

Urbanization is one of the major trends guiding and affecting development both in cities, as well as in surrounding regions and rural areas. Therefore, finding sustainable strategies for cities of all sizes is a fundamental challenge globally, but the focus on small and medium-sized cities allows a better understanding of the changing relationship between urban and rural realities. There is ample research on cities and sustainability, but, unlike most of the other contributions, this book brings culture in sustainable urban development to the core: it investigates, in particular, how participatory culture and artistic expression contribute to the sustainable future(s) of the cities. By focusing on various strategies, policies, and bottom-up practices of small and medium-sized cities in Europe, the book offers a perspective to imagine sustainable future(s) for smaller European cities as well as others worldwide.

Ultimately, the book seeks to explore different models of local sustainability that incorporate emotions and attachment to one's living place and move beyond profit-driven approaches, which tend to use culture and cultural policies only as a means for embellishment of the urban environment and propelling economic viability. The book also implicitly reveals path-dependencies of the development of European cities; each city has its history, which affects its future development. At the same time, cities are not "trapped" in their pasts, and, as the cases in this book show, the human agency, imagination, and cooperative efforts of residents can alter local development trajectories. Therefore, as the editors of the book wisely note, there are no copy-paste strategies. Rather, each city should find its own path to sustainability through innovative re-imagination of its own cultural assets and traditions and through creatively applying the lessons learned elsewhere.

Katriina Soini and Joost Dessein

Figures and tables

Figures

Tables

Notes on contributors

Editorial board – The shaping and development of this book benefitted from thoughtful advice received from an editorial board consisting of Prof. Mariusz Czepczyński (University of Gdansk, Poland), Prof. Anita Kangas (University of Jyväskylä, Finland), Prof. Miloslav Lapka (University of South Bohemia and Charles University, Czech Republic), and Dr. Rolf Hugoson (Umeå University, Sweden).

Helmut K. Anheier, PhD, is President and Dean, and Professor of Sociology at the Hertie School of Governance, Berlin, Germany. He also holds a chair in sociology at Heidelberg University and serves as Academic Director of its Center for Social Investment. He is currently researching the role of foundations in civil society. Helmut Anheier is author of over three hundred publications, and he has won various international prizes for his scholarship. His most recent publication is the second edition of his textbook, *Nonprofit Organizations: Theory, Management, Policy* (Routledge, 2005 and 2014).

Rachel Blanche is Lecturer in Cultural Management at Queen Margaret University, Edinburgh, Scotland. She specializes in cultural policy, having contributed directly to cultural policy formation in Scotland through consultancy work with the Scottish government, the Scottish Arts Council, Creative Scotland, and many organizations across the arts and cultural sectors. Rachel remains active in the field of national and international cultural policy formation and currently serves as a member of the United Kingdom National Commission to UNESCO – Scotland Committee.

Aleksandar Brkić, PhD, is a scholar in the fields of arts management and cultural policy. He is a lecturer at LASALLE College of the Arts, Singapore, as well as a guest lecturer at the University of Arts in Belgrade and Faculty of Technical Sciences in Novi Sad, Serbia. In 2011, he was the recipient of the Cultural Policy Research Award, and his book *Cultural Policy Frameworks (Re)constructing National and Supranational Identities: The Balkans and the European Union* was published in 2013 by the European Cultural Foundation. He works as a creative producer in the intersections of performing arts, visual arts, and design.

Eva Cudlínová is Associate Professor at the University of South Bohemia, Faculty of Economics, České Budějovice, in the Czech Republic. She is a lecturer on ecological and environmental economics. She was educated at the University of Economics in Prague, and previously employed at the Institute of Systems Biology and Ecology, Czech Academy of Sciences. Among her main fields of interest are problems of sustainable development, resource management, economic methods of valuing nature, and global environmental problems from an economic perspective. She is author or coauthor of about sixty articles in scientific journals and dozens of book chapters.

Milena Dragićević Šešić, PhD, is Head of the UNESCO Chair in Cultural Policy and Management (Interculturalism and Mediation in the Balkans) and former President of the University of Arts, Belgrade, Serbia. She is a board member of the European Diploma in Cultural Project Management (Brussels) and the Cultural Policy Research Award (European Cultural Foundation, Amsterdam), and UNESCO expert for a Technical Assistance Mission (Cambodia). She holds Commandeur dans l'Ordre des Palmes Academiques (France) and is a Member of the National Council for Science (2006–2010). She is a guest lecturer at numerous universities and has published fifteen books – including *Art Management in Turbulent Times: Adaptable Quality Management*; *Culture: Management, Animation, Marketing*; *Intercultural Mediation in the Balkans*; *Neofolk Culture*; *Art and Alternative*; *Urban Spectacle*; and *Horizons of Reading* – and numerous essays. Her work has been translated into sixteen languages.

Nancy Duxbury, PhD, is Senior Researcher at the Centre for Social Studies, University of Coimbra, Portugal, and Co-coordinator of its Cities, Cultures and Architecture Research Group. Her current research examines culture in local sustainability planning, cultural mapping methodologies, and community engagement through environmental-artistic projects. She is a member of the European Expert Network on Culture and an adjunct professor at the School of Communication, Simon Fraser University, Canada. She is published internationally and is editor of *Animation of Public Space through the Arts: Toward More Sustainable Communities* (2013) and coeditor of *Cultural Mapping as Cultural Inquiry* (2015). She cofounded and was Director of Research of the Creative City Network of Canada.

Nelly van der Geest, PhD, is a sociologist, theatre maker, and expert in organizational change. At the University of the Arts Utrecht (HKU), Netherlands, she leads the knowledge programme "Art Education in Context" at the Center of Expertise on Education. She publishes on creative partnerships, creative ability as purpose in learning, and the role of artists and designers in participatory and sustainable projects. She lectures in HKU's master's programme, Art Education, on innovation of education due to societal dynamics such as sustainability or migration. In the Dutch national programme, Leadership in Culture (LinC), she leads the "interventure" track, in which upcoming leaders in culture research the adventure of innovation.

Loreta Georgievska-Jakovleva, PhD, is Full Professor of Cultural Studies at the Saints Cyril and Methodius University in Skopje, Republic of Macedonia. She is Editor-in-Chief of the journal *Culture/Култура*, and coeditor of the book *Creative Economies: New Knowledge for New Initiatives* (2014). During the last five years, she has published three books: *Literature and Cultural Transition* (2008); *Identities* (2012); *and Culture and Media* (2014). Her research interests include the areas of identity, cultural memory, cultural policy, and cultural industries.

Michael Hoelscher, PhD, is Senior Lecturer at the Institute of Sociology, University of Heidelberg, Germany. He holds a PhD from the Free University Berlin, and a Habilitation from the University of Heidelberg. From 2006 to 2008, he worked at the University of Oxford, where he still is Associate Research Fellow at the Centre on Skills, Knowledge and Organisational Performance (SKOPE). His main fields of interest are globalization; European integration; cultural, economic, and urban sociology; and higher education. Currently, he is working on a comparative project on "New Inequalities in Creative Cities." His publications include *Cities, Cultural Policy and Governance* (as guest editor with Helmut K. Anheier and Raj Isar) and "Innovation and Creativity" (in the *Encyclopedia of Global Studies*, edited by H. K. Anheier and M. Juergensmeyer, 2012).

Svetlana Hristova, PhD, is an urban sociologist, researcher, and lecturer; Associate Professor at the South-West University, Bulgaria; and the initiator and Chair of the Thematic Area "Urban Management and Cultural Policies of the City" within the European Network of Cultural Administration Training Centers (ENCATC). Her research interests lie in the crossroads of sociology of culture, urban studies, and visual anthropology, as exemplified in numerous projects and publications in English, French, German, Polish, Finnish, and Lithuanian, besides her native Bulgarian. She is author of the books *Border Identity of the Small Town* (2001) and *City: Images, Symbols, Identities* (2002), and is coeditor of *Public Places of European Cities* (2008) and *Public Space: Between Re-imagination and Occupation* (2015).

Rolf Hugoson, PhD, is Senior Lecturer in Political Science at Umeå University, Sweden. He is also a researcher on urban history at the multidisciplinary research centre for regional science, CERUM. He is participating in the research platform "Culture and Cities," focusing on European Capitals of Culture. Hugoson has published two books and a number of articles and chapters on cultural policy, political and rhetoric theory, and diplomatic history. A book in Swedish on the 1950–2010 history of the 2014 European Capital of Culture Umeå is forthcoming in 2015.

Jenny Johannisson, PhD, is Associate Professor at the Centre for Cultural Policy Research, Swedish School of Library and Information Science, University of Borås, Sweden. Her main research interests are local and regional cultural policies understood against the backdrop of globalization processes, and she has published extensively on these issues. Johannisson chairs the scientific committee of the International Conference on Cultural Policy Research (ICCPR), a biennial research conference which is organized in association with the *International Journal of Cultural Policy* (Routledge), and she was coeditor (2009–2012) of the open access scholarly journal *Culture Unbound: Journal on Current Cultural Research*.

Anita Kangas, Professor, PhD, is the Director of the Unit of Cultural Policy/Politics at the University of Jyväskylä, Finland. Her publications and research interests include theory and history of cultural policy, models of local cultural policy, cultural participation, cultural sustainability, and action research methodology. She is a member of the European Expert Network on Culture and the ENCATC Cultural Policy and Cultural Management Research Award international jury, and she has conducted many research projects (for example, Local Government Culture and Sport Services – Access and Democracy; Cultural Sustainability in Rural Areas; History of Cultural Policy; and Construction and Reconstruction in Authorship in the Arts).

Charles Landry is an international authority on the use of imagination and creativity in urban change. He invented the concept of the creative city in the late 1980s, which became a global movement and changed the way cities thought about their capabilities and resources. He has written several books, including *The Fragile City & the Risk Nexus*, with Tom Burke; *The Sensory Landscape of Cities*; and *The Origins and Futures of the Creative City and Culture & Commerce*. He is best known for *The Creative City: A Toolkit for Urban Innovators* (2000); *The Art of City Making* (2006); and *The Intercultural City: Planning for Diversity Advantage*, with Phil Wood.

Miloslav Lapka, PhD, has held several leading positions in the Academy of Sciences of the Czech Republic. Now he is working in the Faculty of Economics, University of South Bohemia, České Budějovice, Czech Republic; and the Faculty of Art, Charles University, Prague. His research interests concern landscape ecology, rural sociology, problems of the human dimension in landscape and global environmental changes, and problems of interdisciplinarity and relationships between sciences and social sciences. Miloslav Lapka is author or coauthor of more than one hundred scientific papers presented in national and international journals as well as chapters and seven books.

François Matarasso is a UK-based freelance writer and researcher with a thirty-five-year career in community arts. He has worked in many European countries, and his work is widely published and translated. His books include *The English Castle* (1993), *Use or Ornament? The Social Impact of Participation in the Arts* (1997), *Only Connect, Arts Touring and Rural Communities* (2004), *Where We Dream* (2012), *Winter Fires, Art and Agency in Old Age* (2012), and *Bread and Salt: Stories of Art and Migration* (2013). For more information see: http://parliamentofdreams.com

Julija Matejić, MA, is a researcher and the Coordinator at the UNESCO Chair in Cultural Policy and Management (Interculturalism and Mediation in the Balkans), University of Arts in Belgrade, Serbia. Following her career as a pianist, she has developed into a professional in the field of arts management. Currently, she is a PhD candidate in the Department of Theory of Arts and Media, University of Arts, on the topic *Theory of Performing Practices in Music (Performativity of Pianism)*. Her research interests also include art in public space, post-memory, and the concept of sustainable development in the field of culture.

Aldo Milohnić, PhD, is Assistant Professor of the History of Theatre at the University of Ljubljana's Academy for Theatre, Radio, Film and Television, in Slovenia. He is Editor-in-Chief of the *Politike* book series, editor or coeditor of numerous anthologies of texts and special issues of cultural journals, and author of the book *Theories of Contemporary Theatre and Performance* (2009). Over the last twenty years, he has been involved in many research projects dealing with the history and theory of performing arts, sociology of culture, and cultural policy issues.

Mišel Pavlovski, PhD, is Full Professor of Cultural Studies at the Saints Cyril and Methodius University, Skopje, Republic of Macedonia. His research interests include cultural studies, postcolonial studies, media studies, and theatre studies. He is the Head Researcher of "The Macedonian Literature Data Base" and was Editor-in-Chief of the international journal *Spektar* (2010–2012). He has published the following monographs: *From an Original Idea to the Creation of a State* (coauthored with Jovan Pavlovski, 1993); *Macedonia: Yesterday and Today* (coauthored with Jovan Pavlovski, 1996); *Pure Play: Biomechanics in the Theatre System of V. E. Meyerhold* (1998); *Theatre and Myth* (2005); and *Creative Economies: New Knowledge for New Initiatives* (coedited with Loreta Georgievska-Jakovleva, 2014).

Katarzyna Plebańczyk, PhD (management in the humanistic sciences), works as an Adjunct Professor at the Institute of Culture at Jagiellonian University, Cracow, Poland. She is the author of several development strategies for cultural institutions that actively promote the idea of mutual understanding between the sender and the recipient of cultural activities. Her research interests encompass a broad spectrum of cultural management topics, including cultural economics and cultural management (e.g. sustainability of cultural organizations, audience development, cultural and food tourism), knowledge management (e.g. the role of cultural observatories), and the strategic management of culture.

Davide Ponzini, PhD, is Assistant Professor of Urban Planning at Politecnico di Milano, Italy. His research focuses on planning theory and urban and cultural policy, and he has extensively published in international journals and books. He is coauthor (with Pier Carlo Palermo) of the books *Spatial Planning and Urban Development* (Springer, 2010) and *Place-Making and Urban Development: New*

Challenges for Planning and Design (Routledge, 2015); and coauthor (with the photographer Michele Nastasi) of the book *Starchitecture: Scenes, Actors and Spectacles in Contemporary Cities* (Allemandi, 2011).

Sakarias Sokka, PhD, has been researching and teaching in the Unit of Cultural Policy/Politics at the University of Jyväskylä, Finland, since 2004. His research interests belong to the field of historical sociology, and his scientific publications, both in Finland and internationally, consider the historical development of Finnish cultural policy and the justification and implementation of public cultural policies.

David Stevenson is Programme Leader for the MA Arts, Festival and Cultural Management at Queen Margaret University, Edinburgh, Scotland. His research is primarily qualitative, and he is particularly interested in interpretative methodologies as approaches that have the potential to give a greater insight into the inherently complex and pluralist landscape in which policy is destined to function.

Elisabete Caldeira Neto Tomaz is a PhD candidate in sociology at the Centre for Research and Studies in Sociology, University Institute of Lisbon (CIES-IUL), Portugal. She holds a bachelor's degree in communication design and a master's degree in communication sciences. Tomaz has participated in many international conferences and has submitted several articles for publication in scholarly journals. She has collaborated with INTELI, a think tank in Lisbon, in several European and national projects and networks related with development policies and cultural/creative policies. She is a member of the COST Action "Investigating Cultural Sustainability."

Introduction

Svetlana Hristova, Milena Dragićević Šešić, and Nancy Duxbury

If our time is labelled the "urban age," it is not simply because more than half of the world's population lives in cities but, more importantly, because "our world order comes to be built on cities and their economies rather than nations and their armies" (Khanna 2010). In the global constellation of cities, scale, positioning, vision, and ambition matter. Responding to critical aggravations in economy, ecology, and equity, cities today are acting as autonomous players seeking original solutions to local, national, supranational, and global problems. However, European small cities are facing diverse challenges besides global economic pressures and environmental threats, suffering from structural disproportions due to the demographic crises of quickly aging populations and workforces fleeing to the more promising job markets of metropolises (EEA 2009). During the recent economic crisis, European *small and medium-sized cities* (SMCs) have been challenged to demonstrate their strengths and capacities. It is precisely at this junction that they have displayed new levels of creative civic imagination in their search for specific innovative solutions within the competition for resources on both European and global stages and to find locally resonant sustainable development solutions.

This book was developed to energize and deepen the discussion about how SMCs are creatively addressing with the means of culture the challenges and risks of the contemporary neoliberal world (Lorentzen and van Heur 2012; Bell and Jayne 2006). Although SMCs represent over 60 percent of European urban settlements and give shelter to the majority of the European urban population, they receive disproportionally less attention in scientific analyses, which traditionally focus on global cities and metropolises. This volume addresses both a deficit of attention to the role of small and medium-sized cities in European sustainable development and an underestimation of the role of culture, artistic expression, and creativity for integrated development as a prerequisite for urban sustainability.

Characterized by interdisciplinary richness and geographical diversity, the book presents a broad collection of case studies and good practices from throughout Europe that illustrate how culture contributes to more sustainable models of city making focused on quality of life, local identities, and creating places where people like to live. The selected variety of regionally specific cultural approaches underline also how local cultural sectors have to be sustainable themselves, in order to operate as engines of integral city development. All these research issues are aspects of the overarching question: what are the roles of culture in sustaining small and medium-sized cities?

On the other hand, although culture is based on traditions, providing continuity "precipitated" into both material and immaterial heritage, still it is not a book about heritage. When the cases in this volume address issues of heritage, it is from the perspective of its contemporary use, reuse, and/or adaptation as a "natural" part of the cultural assets of communities and their places (Fairclough et al. 2015). The book also investigates how participatory culture, community arts, and, more generally, the creativity of civic imagination can contribute to the goal of sustainable futures for SMCs. Ultimately, the book is about European citizens, making their lives through sustaining their cities. In a broader perspective, the cases reveal the intricate and fragile relationships between cultural strategies, economic policies, sustainable development goals, and the national and regional contexts in which cities are embedded. As both Matarasso and Tomaz (in this volume) keenly observe, policy discourses often advocate an economic rationale for cultural strategies, while a community development focus depends largely on actors' power relations and their operating/national frameworks.

From a theoretical point of view, the book represents a conceptual node fusing three dimensions – insights from urban studies concerning small and medium-sized cities, the pressing agenda of sustainability, and cultural policy solutions and recommendations. Within this thematic intersection, the book addresses key issues such as:

- how to move beyond the dominating understanding of sustainability as an ecology-economy-equity balance to a more holistic sociocultural and human-needs-led development conception;
- how to address the contradiction between the profit-driven economic processes of regeneration, gentrification, and commercialization of cities and the emergence of more participatory bottom-up urban cultural movements (i.e. the right to the city);
- how to reconcile the potential contradiction between organizational fields of culture, sustainability, and innovation, which sometimes develops into open tension; and
- how to understand the various ways cities have been "Europeanized" in different national contexts, from simple copy-paste practices to the creative adaptation of European methodologies and innovative applied approaches.

Key concepts and cross-linking themes

The idea of Europolis

Despite the widely spread opinion that many European SMCs have become no more than suburban commuter satellites to the large cities and conurbation urban regions, the examples in this book reveal another reality: places *sensitive* to their environments, which contribute to a high quality of life, aim to be "safe and inclusive, well planned, built and run," and offer "equality of opportunity and good services for all" (EU 2007). *Europolis* is conceptualized as the emergence of a European model – or models, since it incorporates a variety of local approaches – for urban sustainability involving processes of reimaging and cocreating new city futures explicitly using culture. The term *Europolis* is also used to emphasize that these emerging models may possibly be European-specific approaches to sustainable urban development, in contrast to the development of metropolises and large urban agglomerations as a worldwide practice. *Europolis* is not a utopian concept, but it suggests a need to discern the counterbalancing processes enabled by new urban policies to meet economic, social, and ecological challenges, with sensitivity to the agency of participative culture.

Sustainable urban development – the sustainable city

In this book, the concept of *sustainable city* crystallizes as a consolidation of different coextensive lines in existing interpretations of *urban sustainability*. Understood as an issue of new urban culture with social, economic, institutional, technological, and ecological aspects, the book critically analyses the concept of *sustainable cities* from a sociocultural point of view, pleading for an integral holistic approach. This multidimensionality of urban sustainability bears implicit controversies. One of these is set around the concepts of *sustainability* and *the sustainable development of cities*. The concept of *sustainable city* incorporates the idea of smart space organization, taking account of land use, sanitation, and efficient transport – a kind of technologically perfect, car-free, carbon-neutral, no waste, solar-powered paradise, "full of architectural marvels to house new Guggenheim and Louvre collections in stunning new buildings by Frank Gehry and Jean Nouvel" (Khanna 2010). Cities of this type, newly designed in the non-Western world, may pretend to be sustainable, but they remain artificial assemblages, isolated from the calamities of reality and missing the true social glue: "For these emerging global hubs, modernization does not equal Westernization. . . . Western values like freedom of speech and religion are not part of the bargain" (Khanna 2010). This is certainly not the European model of continuous sustainable growth that aims to attentively manage local natural resources and to provide opportunities for inhabitants to develop their own abilities and collective local potentialities.

In exploring sustainable cities from a cultural point of view, the book suggests at least three interconnected and largely complementary possibilities for consideration: culture-based sustainably developing places, livable places, and ecologically sensitive culturized places (cf. Young 2008). The concept of *culture-based sustainably developing places* serves to introduce the necessity for creatively inclusive neighbourhoods or other types of territorial units with a potential for an evolving *sense of place*, which create their own developmental logic within the city itself but often go beyond city "walls" by expanding their regional, national, and international networks, thus offering more opportunities and cultural services for both citizens and visitors (Kangas and Sokka; Matarasso; Johannisson; Plebańzcyk; van der Geest). *Livable places* emphasize as their primary "task" ensuring a high quality of life "no longer defined by high material consumption . . . [but] when people can live a healthy, pleasant and safe life" (EEA 2009, 43). In other words, they can enjoy life on two interlinked levels: the satisfaction of everyday life needs and the fulfillment of life-long goals, which collectively create their sense of meaning (Georgievska-Jakovleva and Pavlovski; Dragićević Šešić, Brkić, and Matejić). *Ecologically sensitive culturized places* put greater emphasis on the revitalization of social fabrics in place-based contexts through renewed relationships with the local natural environment and new commitments to places of cohabitation as reimagined aesthetized places of restored cultural traditions and collective memory (Lapka and Cudlínová; Milohnić).

Civic culture and civic urbanity

While urban regeneration research typically focuses on economic dimensions and impacts, the book offers three important layers of debate that are investigating the potential creativity of community in contexts of participatory openness:

- *Civic imagination*, fostered through participatory creative processes, is a necessary prerequisite for the creation of diversified models of sustainable SMCs, which are tightly linked to promoting a specific local identity on global competitive stages, where metropolitan cities are dominating with institutionalized and financially supported "imagination" through numerous cultural and entertainment organizations. When a city or its civil society offers ongoing, context-specific platforms for expression, *civic imagination* can be an inexhaustible source of continuous innovation and experimentation, contributing to the further valorization of SMCs on a global scale.
- *Civic urbanity* refers to processes of creating connection and purpose (Landry) through being a full citizen, meaningfully engaged with his/her city. Culture is an important part of civic engagement as the authors of this book argue: culture is a public good, and a city's public space with its various cultural functions is a crucial element of sustainable city making. Thus, cultural initiatives are regarded as an essential part of civic urbanity, especially taking into account their social relevancy, capacity to contribute to social inclusion, and ability to regenerate a sense of civic pride.
- *Civic–public partnerships* involve civic actors and movements in dialogue with local governments. Through these partnerships, public values of culture are rearticulated and reemphasized. Partnering with civil society in reconceptualizing and creating broader cultural offers is the most important approach for cities wishing to innovate the scope of activities and assure the well-being of their citizens. Cooperation with different partners from civil society (artists, activists, NGOs, experts, and concerned citizens) allows for more prosperous, just, and inclusive societies, creating a city that is more vibrant

and confident in its own developmental capacity. Through partnerships, both sides are empowered and strengthened, so that civil society does not hesitate to offer its help to local government, and local government becomes closer to its citizens and their needs.

Public sphere: culture in the public interest as a public good

As contemporary society has shifted towards neoliberal market regulation, many people have suffered the throes of, first, the industrial/postindustrial crises and, then, the financial/banking crisis. Some authors claim that what we observe today is a cyclic structural crisis of a profit-driven neoliberal economy. As noted in the Brundtland Report, "economic development is unsustainable if it increases vulnerability to crises" (WCED 1987, 38). Although a new sociocultural paradigm is not yet fully formed, public actions are now set within a generalized but strong disbelief in neoliberal solutions, and this situation has led people to go back to something more "stable," generally characterized by solidarity, mutual support, and redistribution of profits and losses. This process of new urban activism, *urbactivism* (Hristova), has been embraced by bottom-up *artivism* (Dragićević Šešić, Brkić, and Matejić); energized by civic imagination (e.g. *Aalborg Charter of European Cities and Towns Towards Sustainability* [*Aalborg Charter* 1994]; European Sustainable Cities Platform 2013); and reinforced by top-down European policies through a series of public documents (e.g., *Leipzig Charter on Sustainable European Cities* [EU 2007]), supported by considerable financial funding (Duxbury). In this context, culture once again is being considered a public good and an activity in the public interest, with a strong focus on the local space where it is "born" and lives. However, this is not a one-way street: as Landry underlines, civic urbanity develops in the "risk landscape" of cities facing a visible decline in engagement.

As the public sphere has diminished, new movements of urban sustainability have tended to focus on the enhancement and growth of the (sometimes limited) remaining public realm. We observe a return of the importance of the public sphere expressed spatially (public space), virtually (public communication, social media), and with close attention to democracy and public dialogue, enhanced with the means of arts and cultural expression as a public good and cultural right. Cities – local authorities and citizens together – are co-responsible for recreating the public sphere, taking into account its need to develop and support platforms for sociability and debates; public spaces for people to meet and gather; and public media or institutions as spaces for social dialogue. As the public and private spheres become more tightly interlinked, both public and private entities are increasingly expected to play roles in enabling, through partnerships and other arrangements, the openness of the public sphere.

In the global context of cultural policies praising the creative economy and emphasizing its financial contributions to urban development (Vilenica and Kuda.org 2012), this book presents a plurality of approaches and celebrates perspectives where culture is offering different sets of values based on the idea of the public good (necessary for creating conducive conditions for urban sustainability). Contemporary creativity and participatory practices contribute to social inclusion, urban vibrancy and renewal, and the incremental development of meaningful places. In contemporary times when business ideas about efficiency have been transferred into all domains of cultural and civic life, the practices, approaches, and perspectives highlighted in this book illustrate pathways to renegotiating the different values between sectors, accepting the importance of the noncommercial logic of the shared commons (Landry). In this negotiation of values, the importance of culture maintaining its own autonomy is recognized as contributing to public interests (Breznik 2004).

Participatory culture and arts, creative inclusion, and empowerment

Participatory projects are usually self-organized actions that sustain human relations and bonds of trust within a community – they are examples of civic urbanity in action, and they can lead to shared policies (Dragićević Šešić 2006). Acting collectively, these projects not only contribute to the sustainability of a place but also raise civic pride and create collective cultural memories (Assmann and Czaplicka 1995) that are indispensable for sustaining *livable places*. Thus, this book is about *people* making cities sustainable through participatory cultural civic activities. It is also about artistic contributions that inspire and lead many of these actions. As SMCs are usually not highly multicultural (i.e. much less than metropolitan cities), this issue was not often highlighted, but it was not avoided. This is also a book about specificities that shows why SMCs need different approaches and focused investigation, as many of the dominating themes in urban sociology and anthropology do not resonate with SMCs.

Dynamics and tensions

Contemporary cities are battlegrounds of controversial interactions and transversal *tensions* based on different development values and models, which Charles Landry (in this volume) calls *fault lines*: "so deep-seated, intractable and contentious that they shape our entire worldview, such as the conflict between environmental ethics and economic rationality in running countries or cities." This can take the shape of a conflict – for example, between the economic-based gentrification of a city versus the bottom-up "right to the city" movements aiming at community-focused sustainability processes. In many cases, high-level investment policies mean unsustainability for the indigenous inhabitants who cannot afford to continue to live in the upgraded locations. This leads to the disappearance of collective memories and sense of place, and, in the long term, to a new identity for the new inhabitants. The emergence of more participatory processes is the key way to reconcile these different tensions and conflicting interests. Public policies reconceived as developmental strategies *of the community* offer platforms where cultural economics and citizens' interests might go together.

The book structure

Section I – "Culture and sustainable development of European cities: what are the issues?" – is devoted to the main conceptual intricacies of the book. It offers general outlines for understanding the new role of culture in building more sustainable cities in Europe and the special role of small and medium-sized cities in the process of establishing new models of urban sustainability (Anheier and Hoelscher; Landry; Hristova; Milohnić) where culture is both promoter and pattern maker for new values and the engine of new *urbactivism* (Hristova). As mentioned previously, European SMCs have demonstrated strengths and capacities (although limited) in the face of the recent economic recession and have displayed a new level of creative civic imagination as they seek specific innovative solutions within the competition for resources on both European and global stages.

When it comes to urban creativity, it is worthwhile to notice the visions of one of the "fathers" of the *creative city* concept, Charles Landry, now reflecting on *sustainable city making* based on the normative idea of *civic urbanity* framed by the interlinked concepts of a holistic approach, the shared commons, eco-consciousness, healthy urban planning, cultural literacy, inclusivity, intergenerational equity, the aesthetic imperative, creative city making, and an invigorated democracy. Creativity – broadly understood as thinking, planning, and acting with imagination – when applied to (creative) city making, goes beyond the "urban engineering paradigm" and incorporates all aspects of the city: the hardware, the software (i.e. a creative mindset), and the "orgware" (Landry). All this gives rise to so-called *hands-on urbanism* (Milohnić).

In this respect, the four chapters in this section are unanimous: although creative city approaches view culture as an input, throughput, and output factor of urban economies – as part of value chains that include investments and development as well as production, services, distribution, and consumption – culture cannot be viewed only as an instrumental means to sustainability (Anheier and Hoelscher). As Aldo Milohnić put it, culture is not only the number of tickets sold; in essence, culture's impact resides in its ability to (re)produce and/or to challenge the norm as well as its ability to contribute to the well-being of national and local communities. The two case studies of urban gardening in Slovenia aptly reveal the intrinsic art of growing the city as "sharing the management of a space and various participatory processes" in which not only plants are cultivated, but also (or even more importantly) social relations.

This is also the point of departure to understanding the authentic spirit of *Europolis*, which from the very beginning – with its socio-spatial and cultural arrangements and with its tradition of openness and democracy – is essentially European (Hristova; Landry). In today's fluid world, urban society often suffers from a lack of commitment to the city and changing meanings of citizenship. Still, "citizens at their best are makers, shapers, and cocreators of their evolving city," and, as Landry notes, ultimately, with its universal claims, *Europolis is Cosmopolis*.

The authors in this section advocate for a holistic, integral understanding of urban sustainability, a field that earnestly requires better knowledge about how cultural strategies work in this context, so that cities do not waste resources in unsustainable projects. In this conceptual and operational setting, cultural policy has to incorporate sustainability as a key aim in its strategies (Anheier and Hoelscher).

Section II – "Europolis as a project: envisioning more sustainable cities" – illuminates how the integration of culture within sustainable urban development is influenced and mobilized by European policy frameworks and programmes, propelled through intercity relations, and entangled within local regeneration dynamics. The EU policy agenda and funds have spurred on a process of "Europeanization" characterized by policy transferability across political systems, transnational networking, and local-level adaptations. Cities are increasingly encouraged to think holistically and to take an "integrated approach to sustainable urban development," although operational guidance on this process of integration is only now emerging (Duxbury). While the place of culture in integrated and sustainable urban development is being worked out in policy and planning frameworks, within diverse national contexts culture has become embedded in local city imaginaries, urban policies, and the strategic development of many small and medium-sized cities (Tomaz).

Many local authorities looking to renew urban economies and to address social and environmental problems have assumed a proactive, entrepreneurial approach. Cities are increasingly acting as autonomous agents, taking on their own foreign policies and roles through various acts of urban diplomacy and cooperation. Although intercity competition is very high, they are reaching out, regionally and internationally, to learn from and collaborate with other cities, providing heightened visibility, recognition, and other "home advantages." In this context, the European Capitals of Culture programme forms a powerful frame and opportunity for investment, recognition, and promotion of a city and a "launching pad" for local action and change in cities of all sizes, with growing attention to SMCs (Hugoson). Among adjacent cities, we also observe a new stage of urban cooperation to be recognized globally, but one that is dependent on continued economic vitality to fuel, with diminished dynamism and momentum in the context of the crisis (Ponzini).

In the emerging Culture 3.0 context (Sacco 2011), many countries and cities are about to enter a new phase of culture prosumption, when potentially everybody becomes a cultural producer and gains access to new interactive technologies. While not all cities try to enhance their roles as cultural centres, if such engagement occurs, as Hugoson argues, we should expect it to be directed outwards, towards wider flows. This imperative to continually balance local and global dynamics within changing cultural, urban, and political flows is particularly pronounced for SMCs that have suffered processes of deindustrialization and need to reimagine their sustainable futures.

The cases presented in this section show that fully integrating culture in urban sustainable development remains challenging. While some SMCs have developed strategies and forms of governance and coordination involving private, public, and nonprofit actors (Tomaz), economic rationales typically dominate urban development strategies. Faced with the ambiguities and uncertainty of new development tools and weighted down by "dramatic retrenching" of public administration and other limitations imposed by the economic crisis, culture-based economic-dominated renewal strategies and facility development initiatives that are not integrated into broader urban planning are beginning to show their weaknesses (Ponzini). Sustainable urban development requires multidimensional and holistic thinking, integrated planning approaches, and "embeddedness" in local capacities and aspirations. The integration of approaches stressing community engagement and development through cultural participation can determine local consensus, capacity, and the longer-term sustainability of these strategies. The more an initiative manages to engage different groups within and outside the city, the more opportunities there will be for people to be makers of its continuous transformation.

Section III – "Culture for sustainable development in urban policies and practices" – underlines the need for a new strategic option to reconsider the actual questions of urban and regional territorial planning. It advocates a necessity to link expert knowledge with civic engagement, recognizing SMCs as autonomous cultural spaces with their own sociopolitical complexities. In Europe, local governments' culture-led developmental policies use numerous strategies to make their communities sustainable, where economic growth does not dominate but harvests on the sense and meaning of the place. The key issues revolve around access, inclusion, and participation: from access to policy-making processes as new models of social encounters (Kangas and Sokka), through participation in both creating and implementing community-based social, economic, and cultural practices within urban regeneration planning (Johannisson), to developing sets of specific strategies in tune with local aspirations and capacities (Plebańczyk). The case study of the town of Huntly in Scotland is paradigmatic for its policy efforts reimagining the whole town as a venue, where cultural interventions are "closely connected to the lived realities of the populace and rooted in the *culture-as-way-of-life*," an approach that has incrementally transformed the town through imagination and an understanding of the potential of creativity (Stevenson and Blanche).

Examples from Polish, British, and Finnish cities, and deep case study analyses of the Swedish city district of Norrby and of the Scottish town of Huntly, provide a perfect overview of possibilities for policy makers and planners, private corporations and agencies, civil society movements, and other social agents of change to conceptualize and lead together endogenous processes that enable a variety of strategies for each local community. Such movements are providing a chance for residents from the societal margins to find their place through "creative inclusion" (Matarasso) and to link not only top-down forces of urbanization and bottom-up movements and networks but also enclaves of excellence and richness with their poorest counterparts. There is always a danger that "The consequence, whether intended or not, is to build a creative ghetto that parallels the educational, housing, health, and employment ghettos already provided for the poorest" (Matarasso), but this is more likely to happen in metropolises. In SMCs, the convergence of social, cultural, economic, and environmental effects from systemic policies created in "radical proximity" (Cruz 2012) of all social agents can really enable a stronger, more resourceful, and confident city, both cohesive and open at the same time.

Section IV – "Making the city resilient: building communities through artivism" – is devoted to civic urbanity in action: analyzing different sociocultural movements, artists' initiatives, and even art-education platforms involving innovative projects with an *artivistic* approach, which are still rare in academic practices (van der Geest). The chapters in this section aim to illuminate the cultural "drives" that are energizing and moving citizens to become agents of change and responsible actors in raising their claims in the local agenda (Lapka and Cudlínová). They demonstrate how civic imagination becomes an important part in a city's "emotional geography" (Dragićević Šešić, Brkić, and Matejić) and how "sustainable places" become those locations where people want to live and "actively participate in political life through cultural practices" (Georgievska-Jakovleva and Pavlovski).

The studies explore how artists are inspired to become *artivists*, catalyzers of different social processes, and why cultural organizations go beyond their usual and expected roles (centred on representing arts) to stimulate living culture and different forms of social innovation processes, including claims for social justice, well-being, and happiness (Georgievska-Jakovleva and Pavlovski). Even when those actions that reclaim their "right to the city" fail, the process of widening and (re)conquering the lost or diminishing public sphere is continuously developing, making the city more resilient to pressures from both global and local political and economic neoliberal demands. The examples show how responsibility for meaningful development is taken up by civic initiatives, through experimental actions of local autonomous players, inspiring the emergence of *a new civic imagination as a public good* through newly created societal ties of citizens, experts, scientists, activists, and artists who collectively act as "interlocutors across this polarized territory, intervening in the debate itself and mediating new forms of acting and living" (Cruz 2012, 58).

This final section shows an alternative way to develop more open and inclusive cities at the same time, cities that keep neighbourhood traditions but enable reimaging and regenerating derelict spaces and constructing new ones, and that succeed in developing a new conception of *civic culture and civic urbanity*, where free-minded and creative citizens are ready to use their own ideas in the public sphere, launching community actions, programmes, and projects. Small and medium-sized cities become knowledge-creation and dissemination platforms linking the specialized know-how of institutions with the ethical and *in situ* knowledge of communities and creating new *innovative interfaces* between research, planning, artistic intervention, and civic imagination which altogether produce meaningful life for communities in a sustainable city.

Conclusion: how big are the chances of small cities to sustain?

The leading issue of the book, whether and how culture and artistic expression can contribute to sustainable development, gains added value when applied to European SMCs, a third of which are declining and most of which often lack both financial and human resources. The book is intended to invigorate the debate about SMCs and to highlight the emergence of specific European models of sustainable city, *Europolis*, based on creativity and cultural expression by mobilizing civic imagination. European SMCs as autonomous entities in their complex hierarchical spatial linkages, and citizens creating their own places through a thick web of social relations, enable inclusion through different forms of artistic expression, not necessarily highbrow art, but stemming from local traditions and community experience.

The emerging European model of urban development is dependent on human relations, a dimension often forgotten in the big visions of business-focused entrepreneurialism. Civic participation is exemplified in different creative and inclusive practices of artivism that are raising claims for more equality and social justice in urban life. In this context, civic urbanity becomes the leading principle for reinterpreting policies and practices and, especially, for envisioning new paths of human-led (vs. capital-led) sustainable urban development. In Europe, this marks the emergence of a new phenomenon – *urbactivism*, involving both top-down Europe-wide public decisions that lead to the institutionalization and place-embeddedness of the principle of sustainable development, encouraged by numerous financial instruments, and bottom-up social processes based on local artivism and associational mobilization.

This new form of civic urbanity is nourished by bottom-up civic imagination and embedded in the articulation of places, spaces, communities, towns, and cities. It is enabled by a high level of new public awareness and valorization of the symbolic importance of space, and realized through a variety of attachment processes – to place, to nature, to other people, and to heritage. It leads towards a new wave of *urbactivism*, including numerous artivist participatory practices.

Three important areas for future research stem from reoccurring topics in this volume. One is the triple node between *cultural sustainability, creativity, and innovation*, which can be productive but also full of tension. The selected cases offer *different possible answers* to this strategic question: *should culture be sustained or left by itself to be what it is – that is, the formula of vernacular authenticism?* There are smouldering tensions, revealed in the volume, between "global" trends and "European" policies, on one side, and local cultural expressions, on the other, sometimes externalized as conflicts between experts invited from outside and local artists and communities. Different visions coming from diverse local contexts range from full openness and readiness to embrace cultural hybridism, to an explicit rejection of innovations from "outside" and insistence on prioritizing local voices and local ways. Either of these extremes can lead to oversimplifying, missing opportunities of sustainable development: thus, cultural sustainability is reduced to cultural survival, and innovation is perceived as an intrusion from outside. Broader, pluralistic perspectives offer visions of a *fluid* culture-led sustainable development involving processes based on networking and dialogue, contributing to overall developmental success. Such a human-led sustainable development is cast on a careful balance between innovation coming through creative processes and local traditions.

The second issue deserving more research attention is the recognized role of culture as provider of positive emotions and a *cultural atmosphere* that serves as a social glue for local communities. The notion of an *atlas of emotions* would encourage mapping and nurturing spaces that keep the spirit of the place through an interplay of experiences, "structures of feelings," memories, and cultural practices, thus contributing to the sustainability of neighbourhoods. This raises a question about *the new status of place-based emotions in contemporary culture* which remains open: does the cultural "emotivism" trend represent authentic longing for true bonding and belonging, as a response to fears of disorientation in the dissolving structures of the postmodern world, or is it just a more refined hypermodern consumerism where cultural symbolism is playing a central part (Lipovetsky 2006)? Rising cultural sustainability concerns suggest a new perspective to existing contemplations on this issue.

Finally, the volume provides evidence that within the range of the SMCs, there are considerable differences based on their sizes – only the bigger cities have enough financial resources, and this necessitates strategies for local specializations and/or networking. Where urban settlements are too small and weak, their potential lies in networking and joining of resources. There is also a potential for new promising relations between SMC*s and European regions* – based on competition and cooperation simultaneously, which enables new developmental solutions through joint planning and collaborative practices. Among these new democratic governance processes, these can be identified:

- local, regional, national, and international networking as a response to crisis and way of maneuvering into the future;
- intersectoral collaborations;
- public-private and public-civil partnerships; and
- broad civic participation as an important part of new governance processes.

In all cases, a closer look would allow us to elaborate further a more accurate picture about emerging new realities of sustainable European cities, *Europolis*.

The present collection aims to reveal in a pluralistic way emerging models of small and medium-sized European cities as generators of sustainable development and the special contribution of culture to these processes. Contemporary creativity and participatory practices contribute to social inclusion, urban vibrancy and renewal, and the incremental development of meaningful places. An array of case studies highlight how culture has become embedded in local imaginaries, urban policies, and the strategic development plans of SMCs, thus enabling European citizens to take responsibility, equally, for their cities, for their communities, and for their own lives.

References

Aalborg Charter (1994) Available at: www.sustainablecities.eu/fileadmin/content/JOIN/Aalborg_Charter_english_1_.pdf [Accessed 25 January 2014].

Assmann, J., and Czaplicka, J. (1995) Collective memory and cultural identity. *New German Critique*, 65 (Spring–Summer), 125–133.

Bell, D., and Jayne, M. (eds.) (2006) *Small Cities: Urban Experience Beyond the Metropolis*. London: Routledge.

Breznik, M. (2004) *Cultural Revisionism: Culture Between Neo-liberalism and Social Responsibility*. Ljubljana: Mirovni Inštitut.

Cruz, T. (2012) Democratizing urbanization and the search for a new civic imagination. In N. Thompson (ed.), *Living as Form: Socially Engaged Art from 1991–2011* (pp. 56–63). Cambridge, MA: MIT Press.

Dragićević Šešić, M. (2006) Shared policies: The future of cultural development. In B. Cvjeticanin (ed.), *Dynamics of Communication: New Ways and New Actors* (pp. 103–111). Zagreb: Culturelink.

European Environment Agency (EEA) (2009) *Ensuring Quality of Life in Europe's Cities and Towns: Tackling the Environmental Challenges Driven by European and Global Change*. Luxembourg: Office for Official Publications of the European Communities. Available at: www.eea.europa.eu/publications/quality-of-life-in-Europes-cities-and-towns [Accessed 18 October 2014].

European Sustainable Cities Platform (2014) Available at: www.sustainablecities.eu [Accessed 25 January 2014].

European Union (EU) (2007) *Leipzig Charter on Sustainable European Cities*. Available at: www.eukn.org/E_library/Urban_Policy/Leipzig_Charter_on_Sustainable_European_Cities [Accessed 25 January 2014].

Fairclough, G., Dragićcvić Šešić, M., Rogač Mijatović, Lj., Auclair, E., and Soini, K. (2015) The Faro Convention, a new paradigm for socially- and culturally-sustainable heritage action? *Култура/Kultura* (Skopje), no. 8, 1–23.

Khanna, P. (2010, August 6) Beyond city limits: The age of nations is over. The new urban age has begun. *Foreign Policy*. Available at: http://foreignpolicy.com/2010/08/06/beyond-city-limits/ [Accessed 8 May 2014].

Lipovetsky, G. (2006) *Le Bonheur paradoxal. Essai sur la société d'hyperconsommation*. Paris: Gallimard.

Lorentzen, A., and van Heur, B. (eds.) (2012) *Cultural Political Economy of Small Cities*. Abingdon: Routledge.

Sacco, P. L. (2011) *Culture 3.0: A New Perspective for the EU 2014–2020 Structural Funds Programming*. Report for the OMC Working Group on Cultural and Creative Industries. Available at: www.eenc.info/wp-content/uploads/2011/07/pl-sacco_culture-3-0_CCIs-Local-and-Regional-Development_final.pdf [Accessed 24 December 2014].

Vilenica, A., and Kuda.org (eds.) (2012) *On the Ruins of the Creative City / Na ruševinama kreativnog grada*. Novi Sad: KUDA.org.

World Commission on Environment and Development (WCED) (1987) *Our Common Future*. Report of the Brundtland Commission. Oxford: Oxford University Press.

Young, G. (2008) The culturization of planning. *Planning Theory*, 7(1), 71–91.

Section I

Culture and sustainable development of European cities

What are the issues?

1 Cultural sustainability in small and medium-sized cities

What are the issues?

Helmut K. Anheier and Michael Hoelscher

> The 21st century will not be dominated by America or China, Brazil or India, but by the city. In an age that appears increasingly unmanageable, cities rather than states are becoming the islands of governance on which the future world order will be built.
>
> — P. Khanna (2010)

Introduction

Since around 2010, more than half of the world's population has been living in cities. The figure is even larger in industrialized countries with around 72.6 percent for Europe as a whole, 68.8 percent for Eastern Europe, 84.4 percent for Northern Europe, 67.5 percent for Southern Europe, and 77 percent for Western Europe (Hoelscher and Anheier 2012, 366). Therefore, questions regarding the sustainability of cities are highly important (see e.g. UN Habitat 2011). However, the growing body of literature on this issue is focusing on large metropolises while neglecting small and medium-sized cities (SMCs), which are still the home of most people in the world. Henderson and Wang (2007, 288f.), for example, estimate for the year 2000 that nearly half (2.9 billion) of the world's population lived in urban agglomerations and that cities over 100,000 inhabitants accounted for 1.8 billion of these. Within those, 84 percent of cities had below 1 million inhabitants and together accounted for 37.3 percent or 0.67 billion inhabitants. The relative figures are probably even higher for Europe (Hoelscher and Anheier 2012, 363 ff.).[1]

Another shortcoming of the older literature on urban sustainability is its negligence of the role of culture. Only recently has culture emerged as a key element of urban and regional development. Indeed, according to some analysts, cultural policy "has moved from the margins to the very centre" (e.g. Wilks-Heeg and North 2005, 305), and the epigraph by Khanna suggests that cities are becoming the key actors in this field. Cultural policy has become part of how cities compete for resources, be it human capital, innovative businesses, affluent professionals, or tourists. In particular, creative city approaches now frequently shape and inform the discourse, the policies, and the planning of activities for urban development (see Hristova, in this volume; Anheier, Isar, and Hoelscher 2012;

Landry 2000). These approaches view culture as an input, throughput, and output factor of urban economies – as part of value chains that include investments and development as well as production, services, distribution, and consumption. Cultural policy is meant to support such value creation by harvesting creativity, forging and enhancing links among producers, facilitating the formation of creative clusters, and attracting populations with disposable incomes and penchants for high-end consumption – thereby supporting the economic viability of cities that often face industrial decline otherwise.

However, some researchers lament about the growing instrumentalization of culture through unduly strong emphases on its actual and potential contributions to economic development.[2] In their opinion, there is a neglect of intrinsic, artistic, and other essentially noneconomic values taking place, probably at the expense of other dimensions of sustainability and the sustainability of culture itself.

Nevertheless, although research on the role of culture in urban contexts has increased in recent years, there is still a lack of systematic empirical tests of the actual impact of policies and programmes inspired by creative cities and urban sustainability approaches, and especially their contributions to urban development. Given the prominence of creative city debates in urban planning, cultural studies, and policy analysis (e.g. Grodach and Silver 2012), this neglect leaves a glaring gap in knowledge.

In this chapter, we try to tackle and combine the issues of culture, sustainability, and small and medium-sized cities. First, we discuss two different notions of culture (one broad and one narrow conception) and their interplay with sustainability. In the second section, we analyse this topic in more detail with a focus on the specific situation of small and medium-sized cities and the role cultural policy has to play within them. The third section introduces available data sources and proposes some empirical approaches to the analysis of culture and sustainability in SMCs. A short conclusion and outlook for future research ends the chapter.

Two concepts of culture and their interplay with sustainability

When analysing the nexus between culture and sustainability, researchers first have to define both. However, this is not an easy task. Although there are hundreds of definitions of culture,[3] one can distinguish most generally between two approaches to culture. In a narrow sense, culture can be conceived of as "the arts" – that is, what artists create and what is regarded, preserved, exchanged, and consumed as "art." This includes literature, theatre, music, and visual arts and is the focus of much of the conventional cultural policy and even dominates the "creative city" discourse. In contrast, culture in a broad sense is a system of meaning – the social construction, articulation, and reception of meaning and its representation in everyday life. Accordingly, Williams (1967) introduced the concept of culture as "a whole way of life." With regard to different disciplines in the social sciences, we can further distinguish between, for example, social aspects as a system of meaning and values; economic aspects as a system of creation, production, distribution, and consumption; and political aspects as system of power difference, presentation, participation, and other aspects. As this broad concept stresses the process-character of culture, it is probably as, or even more, important for sustainability as the first narrow definition of it.

The term *sustainability* is somewhat seemingly easier to grasp. The UN defines sustainable development as "development that meets the needs of the present without compromising the ability of future generations to meet their own needs" (WCED 1987). With its three pillars – economic demands, environmental resilience, and social equity – this definition is broadly used, though not uncontested. However, the question of how to achieve sustainability in practice is probably as difficult to answer as it is to find a clear definition of culture, and it is addressed in more detail in the other chapters of this book. Instead, we look at some conceptual issues of combining culture and sustainability.

As already mentioned, many scholars as well as practitioners see culture as an important ingredient for the sustainability of a city (see, for example, the elaboration by Hawkes 2001; also European Task Force 1997, 90 ff.). From an economic perspective, many cities, at least in the industrialized countries of Europe and North America, experienced a decline of traditional industrial sectors with the cultural and creative industries seen as new growth sectors (e.g. Sacco 2011). Ecologically, the latter industries also have the advantage of being greener and cleaner industries compared to their predecessors. Social sustainability is secured by the conceived openness and inclusiveness of cultural and creative industries for different population strata, especially for minorities (see Oakley 2012, who is, however, critical of the results in London). These arguments, which dominate the discourse, refer mainly to a narrow concept of culture.

At second glance, we can add some additional arguments on sustainability to the current debate by referring to a wider concept of culture: "Enhancing our cultural competence, our capacity for cooperation and critical thinking, our openness to diversity, our curiosity; all of these are essential to develop a sustainable society where rights, responsibilities and shared well-being can grow and new types of employment can flourish" (Culture Action Europe 2014, 1; see also Hawkes 2001; Nurse 2006). From this perspective, for example, creative destruction (Schumpeter 1965, 83) and creative input for the economy, maybe as a result of changes in the way of living (multigenerational housing, urban gardening, etc.), could serve as a critique of current, unsustainable economic practices. Ecological values, awareness, and consciousness also play a crucial role for sustainable lifestyles. Additionally, the broad acceptance of cultural plurality and diversity (see the 2005 UNESCO Convention [UNESCO 2013]) helps minorities to keep their identities, leads to a decent living, and provides social sustainability, especially in times of globalization.

Culture, both in its narrow and its wider notion, can therefore generate sustainability in cities in many respects.[4] However, there are some caveats. First, a simple and straightforward way of using culture as a tool for sustainability ends is not available, as the interplay between cultural inputs and economic, ecologic, and social outputs and outcomes is extremely complex and not yet fully understood. Second, culture in itself will probably suffer if it is used only for external needs in an instrumental manner. The most important reason for this is that culture, even in the narrow sense, is about meaning, values, and beliefs and has therefore "to make sense" in its own right. As a result, we can put on record that we have to deal with a reciprocal process: culture has to be sustained in itself in the first place to be able to be instrumental for external needs. Third, there is a danger of primarily focusing on the benefits and the sectors connected to a narrow concept of culture, as they are more easily assessed and are also more affirmative of the status quo. However, even the arts need input from a vibrant and diverse culture in the broader sense to be able to fulfill the tasks mentioned.

Taking these caveats into account, cultural policy might be able to generate a beneficial, self-reinforcing process. On the other hand, if cities do not care for this double reciprocal relationship (between culture and sustainability and between the narrow and broad understanding of culture), they might produce an unsustainable vicious circle for both the city and its culture.

Cultural policy in small and medium-sized cities

Cultural policy has gained growing attention during the last few years, and much has already been written on the topic (see, for example, the literature reviews by Anheier, Isar, and Hoelscher 2012; Copic and Srakar 2012). However, much of this research is focused on large metropolises. Therefore, the aim of the following section is to summarize not the current state of the art per se but rather the discussion of some of the issues related more specifically to cultural policy in small and medium-sized European cities with regard to sustainability.

In general, the European city provides on average a comparatively advantageous context for cultural and sustainability policies (see also Landry, in this volume). European cities have a long tradition of autonomy and self-administration. Actually, this is one of the key characteristics of the emergence of cities mentioned by Max Weber (1985), accompanied by a developed public space and more or less strong civil societies (Habermas 1993). In comparison to many other world regions, European cities are relatively rich, have exhibited only minor growth rates during the last decades (or even, posing a new kind of problem, shrinkage), and have relatively less economic and social tensions (low segregation). Many cities are also characterized by well-established cultural and educational institutions as well as historical depth and layering. Some of these traits are especially pronounced in SMCs.

However, with regard to current discussions about cultural policy and the creative city, a clash between these traditional strengths of European cities and new requirements might exist. First, while conventional cultural policy often aimed at and targeted traditional cultural institutions in the narrow sense of the term (e.g. theatres, operas, concert halls, and museums), it now has to broaden its approach by including cultural and creative industries as well as culture in a broader sense of the term. This has led to a shift from an "emergent post-war cultural policy discourse which attempted to intervene against the market in order to secure culture from the miasma of commerciality" to culture as "a potential economic resource"

(O'Connor 2010, 17 and 31). Therefore, cultural policy and cultural actors in the city have to find a new balance within the following set of tensions:

- Is culture mainly seen as an instrument for the integration of citizens or as an advertisement and promotion of a city in its local competition?
- Is culture about preservation of the old (e.g. heritage), or is the emphasis on creativity and innovation?
- Is the focus on strengthening local culture, or is it about attracting global cultural highlights?
- Is the task of culture that of critique or, as in many creative city strategies, of affirmation?

In most of these cases, there is probably no "either/or" but a "both/and," and many of these tensions are as old as the arts and culture itself. Nevertheless, the current strategic inclusion of culture in the discourse on urban renaissance increases the pressure to find satisfying answers.

Second, these new tasks also ask for a new approach of steering away from the old administrative top-down style towards multi-stakeholder governance. While some city administrations still have to learn to cooperate with external actors, in many regions competent partners from civil society acting next to the state and the economy are already available (see, for example, the analysis by Johannisson, in this volume). In the European context, cities are relatively independent from their nation-states in terms of policies; therefore, flexibility for adaptation (though maybe with less resources) exists.

Relating to the above, cultural policy makers in many places adopted the creative city idea. However, the approach is not easily applicable in SMCs. First, the situation in SMCs is very heterogeneous, which means there is no "one size fits all" recipe (see, for example, the German discussion on the distinctiveness or intrinsic logic [*Eigenlogik*] of cities and its critique: Berking 2008, 2012; Kemper 2011). Additionally, a large bulk of the current research and recommendations stemming from the creative city idea are at least implicitly aiming at the large global metropolises. It is therefore often less appropriate for the smaller cities that are in the focus of this book.

Another problem with regard to sustainability is that creative city strategies often focus on large-scale prestigious projects at the expense of smaller and less marketable venues and cultural forms. This might lead to the vicious circle mentioned above. An important task for cultural policy is therefore to balance external claims for and internal demands of culture (see also the tensions mentioned above) Therefore, cultural policy has to:

- secure additional resources stemming from developments in the context of the creative city, but also protect the cultural core against commodification;
- open up traditional cultural institutions towards cultural and creative industries and raise awareness for economic opportunities, but at the same time make sure to not solely let the market rule the cultural game; and

- realize that there are limited resources (currently shrinking in many countries) and that there is a need for selection, without concentrating them too much.

While looking at the potential role of culture for sustainability, a third challenge is that cultural policy has to explicitly incorporate sustainability as a key aim in its strategies.

Although city budgets are shrinking and costs are simultaneously going up, the situation in Europe is still quite comfortable in many respects. Cultural policy and a broad offer of culture are available in many cities, and the data of the European Union's Urban Audit (see Table 1.1) shows that citizens appreciate this. However, Knight (1995) doubts that a creative city strategy is suitable for most SMCs, as they are lacking the resources to compete with larger metropolises in this field, as research has shown huge clustering effects of creative industries (e.g. Oakley 2004).

A strategy for SMCs to nevertheless have a chance of competing with the large global players is to focus on selected fields and analyse their strengths carefully to achieve some kind of sustainability. They cannot afford a broad offer in different fields and cannot compete, with rare exemptions, with the big global players for top attractions. For this reason, cities need comparable information on their relative strengths and weaknesses compared to potential "competitors." Additionally, there is a general need for better knowledge on how and when cultural strategies work in order for cities not to waste resources in unsustainable projects. The next section therefore gives some indicative hints on what data on European SMCs are available and how those can help to generate the necessary knowledge.

Empirical approaches to culture and sustainability in small and medium-sized cities

By addressing the question of culture and sustainability empirically and in a comparative way, we discover the problem that most research focuses on larger cities, while data on SMCs are rare. Instead of giving an overview of some of the more important available studies, the following section seeks to find a solution to circumvent the data problem on SMCs.

Data on specific cities

Compiling data on cities is very resource-intensive, as much data (or at least its presentation/provision) is lacking comparability. For example, a survey comparing cultural spending of cities even in one country (here, Switzerland) encountered huge difficulties in terms of comparable statistics/reporting: the distribution of cultural expenditures across different administrative budgets (education, city development, culture, sports, family, etc.) varied greatly for the cities compared.[5] It is therefore often useful to draw on existing surveys and other data sources even if secondary analysis means that one has to rely on a limited set of indicators.[6]

One of the most important sources for data on SMCs is the European Union's Urban Audit (2014) project. After some pilot surveys in the 1990s, data on cities in all EU member states and in Switzerland, Norway, and Turkey (and more recently in Croatia and Iceland) have been collected every three years since 2004. The thematic domains covered by the Urban Audit data are outlined in Table 1.1.

In the last wave, data (overall, more than two hundred variables) were collected for nearly one thousand European cities with more than fifty thousand inhabitants, allowing for a wide range of analyses (see Eurostat 2012 for additional information). However, data on cultural aspects are rare to nonexistent. Therefore, the Urban Audit is supplemented by a survey on "Quality of live in cities: Perception survey in 79 European cities [and four surrounding areas]" (European Commission 2013b). Although many of the cities in the survey are larger than five hundred thousand, more than half (forty-three) are smaller (labelled as small, medium, and large in the report; see Table 1.2 and Dijkstra and Poelman 2012). In each city, around five hundred respondents are interviewed about their satisfaction with cultural facilities, city administrative services, and environment, besides other topics.

Cross-national comparisons should be interpreted cautiously, as most countries are represented by just one city/case, but comparisons between larger and smaller cities are possible. Additionally, comparisons over time are also available. A quick look at the data reveals, for example, that citizens of larger cities more often access cultural facilities, while, in contrast, citizens within smaller cities are more content with their environments.

An important finding of the report (European Commission 2013b) is that in all but one city (Valletta) a majority of respondents are satisfied with the cultural facilities. However, there are huge differences between cities and over time. It would be interesting to analyse with more in-depth case studies whether one can identify reasons for the reported sharp falls in Rostock and Newcastle or the

Table 1.1 Thematic domains covered by the Urban Audit

1. Demography	**4. Civic involvement**	**7. Travel and transport**
1.1 Population	4.1 Civic involvement	7.1 Travel patterns
1.2 Nationality	4.2 Local administration	**8. Information society**
1.3 Household structure	**5. Training and training provision**	8.1 Users and infrastructure
2. Social aspects	5.1 Education and training provision	8.2 Local e-government
2.1 Housing	5.2 Educational qualifications	8.3 ICT sector
2.2 Health	**6. Environment**	**9. Culture and recreation**
2.3 Crime	6.1 Climate/Geography	9.1 Culture and recreation
3. Economic aspects	6.2 Air quality and noise	9.2 Tourism
3.1 Labour market	6.3 Water	
3.2 Economic activity	6.4 Waste management	
3.3 Income disparities/ Poverty	6.5 Land use	

Table 1.2 Small and medium-sized cities in the Quality of Life Perception Survey by country

Country/city	*No. of inhabitants older than 15*	*Country/city*	*No. of inhabitants older than 15*
Belgium		**Malta**	
Antwerpen	408 643	Valletta	5 479
Liege	164 178	**Netherlands**	
Bulgaria		Groningen	165 697
Burgas	172 826	**Austria**	
Czech Rep.		Graz	231 347
Ostrava	282 958	**Poland**	
Denmark		Bialystok	255 280
Aalborg	168 728	Gdansk	395 271
København	464 858	**Portugal**	
Germany		Braga	151 827
Leipzig	463 164	Lisboa	477 239
Rostock	181 582	**Romania**	
Estonia		Cluj-Napoca	276 407
Tallinn	336 683	Piatra Neamţ	94 807
Greece		**Slovenia**	
Irakleio	109 358	Ljubljana	236 011
Spain		**Slovakia**	
Malaga	477 216	Bratislava	378 952
Oviedo	199 122	Kosice	199 308
France		**Finland**	
Rennes	311 932	Oulu	119 381
Strasbourg	375 076	**Sweden**	
Italy		Malmö	252 829
Bologna	338 268	**United Kingdom**	
Verona	229 841	Belfast	221 712
Rep. of Cyprus		Cardiff	246 018
Lefkosia	204 179	Glasgow	478 574
Latvia		Manchester	315 244
Riga	423 118	**Iceland**	
Lithuania		Reykjavik	161 857
Vilnius	453 866	**Norway**	
Luxembourg		Oslo	491 181
Luxembourg	86 022	**Switzerland**	
Hungary		Geneva	162 896
Miskolc	156 230	Zurich	329 132

rising satisfaction in Napoli and Burgas. Other interesting results are that the size of a city has a negative impact on satisfaction with environmental issues and that an increasing number of people feel that their city is involved in fighting climate change. Again, case studies could analyse what kinds of campaigns have been the basis for successfully raising awareness.

Other interesting data sources are different index-initiatives such as the Council of Europe's Intercultural City Index (ICI)[7] or the ESPON-project "The Role of Small and Medium-Sized Towns" (SMESTO). While these projects often cover only a limited number of cities (e.g. sixty for the former, fourteen case studies for the latter), they can provide deeper insights on specific issues. For additional analyses, data from different sources with enough overlap could be combined.

The OECD is also providing data on regional policies, but their two databases[8] comprise data only on larger regions (two thousand in thirty-four countries) or on 275 larger metropolitan areas with more than five hundred thousand inhabitants. At least, the former allows contextualizing SMCs in their regional context; the second allows comparisons between SMCs and large cities to a certain extent.

While we argued that it is often, at least for a first approach, useful to draw on existing data sources, the overview of already available surveys and databases has made it clear that evidence on the relationship between culture and sustainability is rather limited and that there are still huge data gaps, calling for joint initiatives to collect additional data.

Survey data without specific city-focus

Another important data source that to our knowledge is underused with regard to city analyses is the increasing number of international population surveys. Many of them have some information on the locality the respondent is coming from. For example, the Eurobarometer surveys feature a variable on the type of community one is living in with the categories of "rural area or village," "small or medium-sized town," and "large town/city." Similar variables are available in the surveys of the International Social Survey Programme (ISSP), the World Values Survey (WVS), and the European Social Survey (EVS). These surveys include a huge variety of variables with regard to the topics of culture and sustainability. Although they do not allow for the identification of specific cities,[9] they nevertheless yield important information on differences between larger and smaller cities, and some surveys even allow comparisons across world regions. From this perspective, although specific cities and the impact of their cultural policies cannot be assessed, rather more general conclusions on relationships between different aspects of sustainability, culture, and city size can be drawn.

To give an easy descriptive example, Figure 1.1 compares cultural activities of citizens across countries and different city sizes (see Table 1.3 for additional data). The data is taken from the Special Eurobarometer 399: "Cultural Access and Participation" (European Commission 2013a).[10]

What becomes apparent is that city size is positively correlated with cultural activities and that this is true for different kinds of activities alike. On average, cinemas are visited more often than theatre or ballet, and cinema culture seems

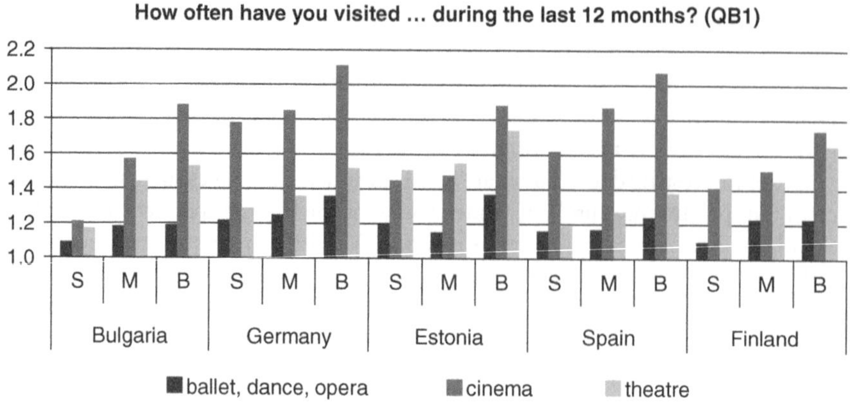

Figure 1.1 Comparison of selected cultural activities across countries and city size (mean of respondents from small [S], medium [M], and big [B] cities; scale ranging from 1 = not in the last 12 months to 4 = more than five times)

to be most developed in Germany and Spain. Besides such easy comparisons, the survey data allows background variables of respondents (such as income, educational background, age, etc.) to be taken into account as well as relationships between different cultural activities or between culture and sustainability issues.

While the analyses can become very powerful based on the large amount of available data and advanced statistical techniques, the already mentioned shortcoming of not being able to identify specific cities and their cultural policies definitely limits the reach of this approach. It is more about general questions of the urban situation instead of tackling what is discussed in Germany and elsewhere under the topic of the intrinsic logic of cities (*Eigenlogik*).

New approaches to combining data meaningfully

A third group of approaches to gain empirical insights from more or less quantitative sources on SMCs is the call to collect and combine data in new and creative ways. Evolving techniques of data mining and "big data" approaches can be used to reap information from very heterogeneous data sources, especially the Internet, complementing the survey data described in the two prior sections. A sustainability-related example would be the analysis of comments on air pollution from visitors in hotel reviews or city-specific blogs. This could even be combined with data from meteorological stations or other "objective" data. Geographical information systems allow aggregating and analysing these data even for smaller parts of cities (see Thierbach et al. 2014 on spatial analyses of social processes more generally).

Researchers have, for example, followed the "global buzz" on the Internet about different large cities, based on Google search counts for one year (GaWC 2009–2010). Similar analyses could be done for SMCs as well. Adding a specific focus to the "buzz" would allow evaluating, for example, the national or global impact of a certain cultural event on the city image.

Another interesting option is a "missing values" approach. Under certain circumstances it is possible to substitute interesting variables (that were not asked in specific city surveys) by data imputation means from other surveys. If, for example, a strong correlation between certain socio-demographic characteristics and cultural preferences can be established on the basis of large-scale surveys, and if we know the distribution of socio-demographic characteristics of a city or its quarters, it might be possible to convey the cultural preferences prevailing in the city.

The Internet, as well as new and powerful data collection (e.g. automated data retrieval from city governments' websites) and analysis tools, will together with the creativity of researchers open up new and increasingly valuable data sources for the analysis of SMCs in the future.

Conclusion

The aim of this chapter is twofold. The first part, based on theoretical considerations and a review of current research, conceptually analysed culture and sustainability in the city context. It started by describing the different aspects or dimensions of the relationship between culture and sustainability from a theoretical point of view. A first conclusion is that a broader notion of culture should be adopted with regard to current research, even if data is not readily available. While there is a growing interest especially in the cultural and creative industries in city contexts, noneconomic aspects of culture, such as values or lifestyles, are at least equally important in relation to the topic of sustainability. Such a broader concept of culture would help to overcome the often existing narrow focus on economic arguments in the debate.

Following from this, a second conclusion is that culture cannot be seen only in an instrumental way as a means to sustainability. Culture – both in its narrow or broader understanding, as transmission of past achievements or as creative force – is inseparably connected to meaning and values. It therefore needs an intrinsic core and has to be valued in its own right. If culture is supposed to benefit sustainability, one first has to make sure that culture is in itself sustainable by supporting it in its own right.

As lamenting research lacunae and shortcomings of existing approaches is, however, inexpensive but not very productive, the second part tried to outline some potential empirical approaches for overcoming data problems with regard to SMCs. It then became clear that the problem is not so much data shortage. A plethora of different data sources exists. However, (a) the data is quite heterogeneous and often not comparable; (b) surveys often cover only a limited number of cities, and overlap across surveys is seldom; and (c) specialized surveys cover only a certain topic, prohibiting analyses across domains such as culture *and* sustainability. All this makes quantitative analyses especially difficult.

To overcome these problems, a combination of qualitative and quantitative approaches (mixed methods) is necessary. Qualitative studies can look at specific cities in detail and provide thick descriptions, but they lack the chance to control for a larger number of background factors or to lead to generalizable results. Quantitative comparative analyses, on the other hand, can take city-specific situations

Table 1.3 Comparison of cultural activities across countries and city-size (full data) (mean of respondents from small [S], medium [M], and big [B] cities; scale ranging from 1 [not in the last 12 months] to 4 [more than five times])

		Ballet, dance, opera	*Cinema*	*Theater*	*Library*	*Museum*	*Reading a book*	*Concert*	*Sightseeing*	*Cultural broadcast on TV/radio*
Bulgaria	S	1.09	1.21	1.17	1.30	1.24	1.89	1.28	1.41	2.58
	M	1.18	1.57	1.44	1.40	1.42	2.41	1.46	1.76	2.78
	B	1.19	1.88	1.53	1.19	1.51	2.63	1.49	1.96	3.12
Germany	S	1.22	1.78	1.29	1.32	1.50	2.54	1.52	1.89	2.64
	M	1.25	1.85	1.36	1.45	1.59	2.84	1.59	2.01	2.77
	B	1.36	2.11	1.52	1.59	1.79	3.01	1.76	2.22	2.86
Estonia	S	1.20	1.45	1.51	2.09	1.41	2.65	1.61	1.69	2.32
	M	1.15	1.48	1.55	1.98	1.50	2.80	1.70	1.74	3.21
	B	1.37	1.88	1.74	2.07	1.80	2.97	1.86	2.13	3.26
Spain	S	1.16	1.62	1.19	1.49	1.34	2.11	1.42	1.64	2.65
	M	1.17	1.87	1.27	1.71	1.38	2.20	1.43	1.80	2.83
	B	1.24	2.07	1.38	1.78	1.63	2.35	1.46	1.97	2.86
Finland	S	1.10	1.41	1.47	2.29	1.19	2.44	1.53	1.22	2.64
	M	1.23	1.51	1.45	2.55	1.46	2.75	1.55	1.49	2.74
	B	1.23	1.74	1.65	2.58	1.71	2.81	1.69	1.88	3.12

Data source: European Commission 2013a.

into account only up to a limited extent but are able to control for additional background variables on the micro-level of the individual, on the meso-level of cities or cultural organizations, and on the macro-level of whole countries. Comparisons of cities of different sizes, within countries, across Europe, and even across different world regions are possible. The combination of both approaches allows tackling the relationship between culture and sustainability from different perspectives and allows the triangulation of the results to inspire and inform evidence-based policy making. While good quantitative data for SMCs is not easily available, some interesting data sources that would allow the extraction of such data as well as some techniques for combining and analysing it were introduced. For example, major population surveys often have, though rough, measures for the size of the urban context of respondents, allowing for simple though large-scale comparative analyses. Additional data can be collected on the Internet by innovative means.

Altogether, substantial-methodological synergies as a result of broadening the concept of culture (theory) and using new data sources (methods) will help to advance the analysis of the relationship between culture and sustainability and provide new insights into this important topic.

Notes

1 For sustainability in the context of urban growth in small and medium-sized cities in developing countries, see Cohen (2006).

2 For example, Hartley (2005) discusses the marginalization and commodification of "art" as result of a growing emphasis on "creativity" (see also O'Connor 2010, 54 ff.).

3 As is well known, Kroeber and Kluckhohn (1952) collected more than 150 definitions in 1952, and the situation has not improved over the last decades. This discussion is already routinized, yet it is growing over the years. The logic of the development of different concepts of culture is well presented in a seminal work of Raymond Williams (1993) and further elaborated by Terry Eagleton (2000).

4 As a result, it is sometimes argued, with good reasons, that culture is a fourth pillar of sustainability (e.g. Hawkes 2001; Nurse 2006). We prefer, at least in the current context, to conceptualize culture as a cross-sectional issue important within all of the other three pillars, as culture in its broad sense helps to define what is sustainable in economic, ecological, and social terms in the first place.

5 This research was presented at the conference of the European Sociological Association (ESA) in Geneva by Luc Gauthier, Karine Darbellay, and André Ducret (2011). The city governments involved did not allow the authors to publish their report.

6 Many reports on cities include indicators on cultural or sustainability topics. See, for example, the study on smart cities by Giffinger et al. (2007).

7 See www.coe.int/t/dg4/cultureheritage/culture/cities/newsletter/newsletter11/newsletter11index_EN.asp.

8 See www.oecd.org/gov/regional-policy/regionalstatisticsandindicators.htm.

9 To a certain extent, such identification might be possible, especially for larger cities, as some of the surveys also include information on the region where the interview was conducted. Combining the size of city with the region might allow certain cities to be identified.

10 As the scale is not metric, taking the average is not really appropriate but is a good approximation.

References

Anheier, H. K., Isar, R., and Hoelscher, M. (eds.) (2012) *Cities, Cultural Policy and Governance*. London: Sage.

Berking, H. (ed.) (2008) *Die Eigenlogik der Städte: Neue Wege für die Stadtforschung* (Vol. 1). Frankfurt [u.a.]: Campus.

Berking, H. (2012) The distinctiveness of cities: Outline of a research programme. *Urban Research and Practice*, 5(3), 316–324.

Cohen, B. (2006) Urbanization in developing countries: Current trends, future projections, and key challenges for sustainability. *Technology in Society*, 28(1–2), 63–80.

Copic, V., and Srakar, A. (2012) *Cultural Governance: A Literature Review*. European Expert Network on Culture (EENC). Available at: www.eenc.info/wp-content/uploads/2012/11/VCopic-ASrakar-cultural-governance-literature-review-final.pdf [Accessed 8 December 2014].

Culture Action Europe (2014) *Culture Action Europe's Appeal for the European Elections 2014*. Available at: www.wearemore.eu/wp-content/uploads/2014/01/CAEs-manifesto-for-the-European-Elections.pdf [Accessed 6 November 2014].

Dijkstra, L., and Poelman, H. (2012) Cities in Europe – the new OECD-EC definition. *Regional Focus* (January). Brussels: European Commission.

Eagleton, T. (2000) *The Idea of Culture*. Malden, MA: Blackwell.

European Commission (2013a) *Cultural Access and Participation* (Special Eurobarometer 399, Wave EB 79.2). Luxembourg: Publications Office of the European Union.

European Commission (2013b) *Quality of Life in Cities: Perception Survey in 79 European Cities* (Flash Eurobarometer 366). Luxembourg: Publications Office of the European Union.

European Task Force on Culture and Democracy (1997) *In from the Margins: A Contribution to the Debate on Culture and Development in Europe*. Strasbourg: Council of Europe Publishing. Available at: www.coe.int/t/dg4/cultureheritage/culture/resources/Publications/InFromTheMargins_EN.pdf [Accessed 6 November 2014].

European Union's Urban Audit (2014). Available at: epp.eurostat.ec.europa.eu/portal/page/portal/region_cities/city_urban [Accessed 12 November 2014].

Eurostat (2012) *Methodological Manual on City Statistics*. Luxembourg: Eurostat.

Gauthier, L., Darbellay, K., and Ducret, A. (2011) *Researching Cultural Policies in Switzerland: Methodological or Political Issues?* Presentation at the European Sociological Association conference, 7–11 September 2011, RN07/Session 6.f.

Giffinger, R., Fertner, C., Kramar, H., Kalasek, R., Pichler-Milanovic, N., and Meijers, E. (2007) *Smart Cities: Ranking of European Medium-Sized Cities*. Vienna: Centre of Regional Science.

Globalization and World Cities Research Network (GaWC) (2009–2010) *Global Buzz: Measuring Short-Term Changes Among Cities in Globalization*. Available at: www.lboro.ac.uk/gawc/projects/projec88.html and www.lboro.ac.uk/gawc/globalbuzz.html [Accessed 6 November 2014].

Grodach, C., and Silver, D. (eds.) (2012) *The Politics of Urban Cultural Policy: Global Perspectives*. London: Routledge.

Habermas, J. (1993) *Strukturwandel der Öffentlichkeit: Untersuchungen zu einer Kategorie der bürgerlichen Gesellschaft* (3. Auflage ed.). Frankfurt am Main: Suhrkamp.

Hartley, J. (2005) Creative industries. In J. Hartley (ed.), *Creative Industries* (pp. 1–40). Malden, MA: Blackwell.

Hawkes, J. (2001) *The Fourth Pillar of Sustainability: Culture's Essential Role in Public Planning*. Melbourne: Common Ground Publishing/Cultural Development Network.

Henderson, J. V., and Wang, H. G. (2007) Urbanization and city growth: The role of institutions. *Regional Science and Urban Economics*, 37(3), 283–313.

Hoelscher, M., and Anheier, H. K. (2012) Indicator suites. In H. K. Anheier, Y. R. Isar, and M. Hoelscher (eds.), *Cities, Cultural Policy and Governance* (pp. 353–442). London: Sage.

Kemper, J. (ed.) (2011) *Lokalistische Stadtforschung, kulturalisierte Städte: Zur Kritik einer "Eigenlogik der Städte."* Münster: Westfälisches Dampfboot.

Khanna, P. (2010, August 6) Beyond city limits: The age of nations is over. The new urban age has begun. *Foreign Policy*. Available at: http://foreignpolicy.com/2010/08/06/beyond-city-limits/ [Accessed 21 July 2014].

Knight, R. V. (1995) Knowledge-based development: Policy and planning implications for cities. *Urban Studies*, 32(2), 225–260.

Kroeber, A. L., and Kluckhohn, C. (1952) *Culture: A Critical Review of Concepts and Definitions*. Cambridge, MA: Peabody Museum of American Archaeology and Ethnology, Harvard University.

Landry, C. (2000) *The Creative City: A Toolkit for Urban Innovators*. London: Earthscan.

Nurse, K. (2006) *Culture as the Fourth Pillar of Sustainable Development*. Report prepared for the Commonwealth Secretariat, London.

Oakley, K. (2004) Not so cool Britannia: The role of the creative industries in economic development. *International Journal of Cultural Studies*, 7(1), 67–77.

Oakley, K. (2012) Rich but divided . . . the politics of cultural policy in London. In H. K. Anheier, R. Isar, and M. Hoelscher (eds.), *Cities, Cultural Policy and Governance* (pp. 204–211). London: Sage.

O'Connor, J. (2010) *The Cultural and Creative Industries: A Review of the Literature*. Newcastle upon Tyne: Creativity, Culture and Education.

Sacco, P. L. (2011) *Culture 3.0: A New Perspective for the EU 2014–2020 Structural Funds Programming*. Report produced for the OMC Working Group on Cultural and Creative Industries on behalf of the European Expert Network on Culture (EENC). Available at: www.eenc.info/wp-content/uploads/2011/07/pl-sacco_culture-3-0_CCIs-Local-and-Regional-Development_final.pdf [Accessed 6 November 2014].

Schumpeter, J. A. (1965) *Capitalism, Socialism and Democracy*. London: Allen & Unwin.

Thierbach, C., Raschke, A. L., Hering, L., and Baur, N. (eds.) (2014) Spatial analysis in the social sciences and humanities. Special issue of *Historical Social Research*. Cologne: GESIS.

UN Habitat (2011) *Cities and Climate Change: Global Report on Human Settlements 2011*. London: Earthscan.

UNESCO (2013) *Basic Texts of the 2005 Convention on the Protection and Promotion of the Diversity of Cultural Expressions* (2013 ed.). Paris: UNESCO.

Weber, M. (1985) Die nichtlegitime Herrschaft (Typologie der Städte). In M. Weber (ed.), *Wirtschaft und Gesellschaft: Grundriss der verstehenden Soziologie* (Fünfte, revidierte Auflage [besorgt von Johannes Winckelmann], Studienausgabe ed., pp. 727–814). Tübingen: J.C.B. Mohr (Paul Siebeck).

Wilks-Heeg, S., and North, P. (2005) Cultural policy and urban regeneration. *Local Economy*, 19(4), 305–311.

Williams, R. (1967) *Culture and Society: 1780–1950*. London [u.a.]: Chatto & Windus [u.a.].

Williams, R. (1993) The idea of culture. In J. McIlroy and S. Westwood (eds.), *Border Country: Raymond Williams in Adult Education* (pp. 57–77). Leicester: National Institute of Adult Continuing Education.

World Commission on Environment and Development (WCED) (1987) *Our Common Future*. Report of the Brundtland Commission. Oxford: Oxford University Press.

2 Civic urbanity

Looking at the city afresh

Charles Landry

Setting the scene

Ten themes shape the dilemmas, challenges, and opportunities for the twenty-first-century city. Each has relevance to how we live and shape our places. They provide an urban narrative I call *civic urbanity* that seeks to contain the explosive mix of centrifugal and centripetal forces we increasingly find in cities. They help reshape how we can rethink urbanity.

The interlinked concepts to rethink urbanity are: holistic thinking, planning, and acting; the shared commons; eco-consciousness; healthy urban planning; cultural literacy; inclusivity; intergenerational equity; the aesthetic imperative; creative city making; and an invigorated democracy. Together they frame the modern idea of civic urbanity. This idea seeks to realign individual desires and self-interest within a collective consciousness that is focused on responsibilities for "us" or "our joint world or city" rather than choices that are only for "me" and my more selfish needs.

Urbanity and being urbane has a proud history. It is important to recapture its best features. The tradition of urbanity, as conventionally understood, is European in its origin. To use a modern term, we can say that it focuses both on "the right to the city" and on "responsibility for the city." Urbanity, as a desirable state of being, first arose in the Italian city-states, especially during the Renaissance, and it then marked the movement towards meritocracy and freeing individuals from the yoke of feudalism. The German phrase *Stadtluft macht frei* (city air makes you free) encapsulates this idea. In time, though, the notion of urbanity degraded, ending with the idea of the *flâneur*, someone who watches urban life go by, but uncommitted to the needs of the collective whole.

The cultural drive

Civic urbanity is a normative idea. It is a promise for a better city. It taps into our deeper yearnings for connection and purpose. It does not come naturally. It has to be fostered and can become part of a new common sense if practiced and encouraged by revised regulations and incentives regimes and programmes.

Or, to put it another way, residents are swayed by the culture of a place so it becomes self-understood behaviour or seems the right thing to do. Since it is not the default position of citizens, urban professionals, or politicians, it is the process or negotiation about why civic urbanity should be taken seriously that is, in itself, *cultural*.

At its core, city making involves arguing about values and making choices based thereon. Then it entails applying values and using politics to turn values into policies and exerting power to get your way. Choices reflect beliefs and attitudes based on values and value judgements. These are shaped by our culture. In this way, the scope, possibilities, style, and tenor of a city's physical look and its social, ecological, and economic development are culturally shaped, and culture moves centre stage. If, for example, a culture invests its faith only in the market principle and trusts the drive of capital to produce sensible choices, the logic, interests, and points of view of those who control the markets will count for more than those who believe market-based decision making is an impoverished theory of choice making.[1] If a culture holds that individual choice is everything – individuals always know best – this impacts the city. Conversely, if people believe the idea of a public, common, or collective good has value and is beyond the vagaries of the market, a different city evolves. Any culture-based argument implies or proposes a trajectory, a plan of action, or even a manifesto of what to do next.

Secular humanism

The ethical frame that best supports civic urbanity is the secular humanist position. This privileges civic values, which in essence seek to foster competent, confident, and engaged citizenship. It is concerned with the capabilities, interests, and achievements of human beings. It does not decry the virtues of science or the sustenance that religion or other belief systems give. In a city context, its aim is to ensure that people of difference live together in relative peace and accord as the city is, in part, defined by and made up of a majority of people who do not know each other (i.e. are strangers). The attempt to arrive at practical standards (that provide principles to guide common views and behaviour and to help resolve conflicts) aims to provide a frame within which difference can be lived and shared with mutual respect.

Secular humanism as a core Enlightenment project has been drained of confidence. It feels exhausted, and consequently it is mistakenly accused of being "wishy-washy" with no apparent point of view. Its confidence needs to be restored. The confident secular humanist view proposes a set of civic values and rules of engagement, which include: providing settings for a continually renewing dialogue across differences, cultures, and conflicts; allowing strongly held beliefs or faiths expression within this core; and acknowledging the "naturalness" of conflict and establishing means and mediation devices to deal with difference. It seeks to consolidate different ways of living, recognising arenas in which we must all live together and those where we can live apart. It generates structured opportunities to learn to know "The Other," to explore and discover similarity and difference. Based on the subsidiarity principle, it wishes to drive decision making downward. This implies much greater decentralization and devolution of power, while central government takes on a more subsidiary role. Altogether, this enhances participation and connectivity at the local level and helps generate interest, concern, and responsibility.

Secular does not mean emotionally barren. In fact, I treasure the heightened registers of being that spirituality evokes. Indeed, its animating force may be just the thing that makes some cities more liveable in than others.

Comprehensive sustainability

The wider notion of sustainability, one that is well beyond merely environmental concerns, embeds itself in this thinking. Taken together, the ten themes capture many of the fault lines, battlegrounds, and strategic dilemmas in city making that need resolving in order to address the disparate needs, urgencies, and crises of the city and the wider world. Fault lines are change processes that are so deep-seated, intractable, and contentious that they shape our entire worldview, such as the conflict between environmental ethics and economic rationality in running countries or cities. Discussions and policy debates around fault lines often become battlegrounds, since opposing opinions are strong, such as in deciding the primacy of holistic or specialist thinking. Some see cities as composed of interacting wholes that are more than simply the sum of the parts, while others look at the fragments within narrow specialisms.

Strategic dilemmas concern the difficult choices that have to be made between two or more alternatives, which are often equally undesirable. Sustainability, as described here, is both the choice-making process and the result of doing something about fault lines, battlegrounds, and dilemmas. Put simply, for instance, if a city allows differences to become hot conflicts or lets inequality lead to disruption or overrigid rules to constrain creativity, the city cannot become resilient to shocks or become sustainable itself.

Europolis to cosmopolis

The foundations of city making as proposed here may have their origins way back in European history, so giving credence to a notion of *Europolis* as a certain type of city. This city is based on a set of underlying principles. Central to this is a sense of the public realm conceived both as a physical place and as a political environment where a discussion culture can develop that generates a self-reinforcing civic culture. This idea in principle has broader resonance across cultures and political systems. For instance, who would argue against being eco-aware or having an effective governance system? Therefore, civic urbanity stakes a claim to playing a part in developing a new cosmopolis inspired by European urbanity.

Ten themes

1. Thinking in the round: holistic thinking, planning, and acting

The starting point is to think in an *integrated and connected* way. Only then can we discern the linkages and dependencies that help us understand the deeper dynamics of cities and how to make the most of their potential. This requires a changed mindset and is difficult to prescribe. Yet, increasingly, decision makers realize that silo thinking and strict departmentalism do not offer the complex solutions we need. We know we need to see the parts and the whole simultaneously.

For centuries we have been splitting knowledge and insight into fragments, boxes, and segregations. From this have grown many inventions and innovations, albeit moving along a narrow furrow. The evaluation of everything from a perspective of specialism and narrowness is a defining characteristic of contemporary society. Narrowness is the managerial paradigm and default mechanism that has embedded itself into how companies, community organizations, and the public sector operate – even while "partnership building" is a mantra of the age. Others prefer to call this narrowness "focus." For the majority, narrowness is the prism through which any activity is judged. Narrowness has its experts, consultants, interest groups, specialist literature, associational structure, and lobbying bodies. It has formalised itself. This has made us lose the art of holistic thinking.

Holism is a scientific theory with a proud history of over one hundred years, but its insights were battered into submission in the race to understand ever-smaller bits of the puzzle. This tradition of thinking has great strengths and is gaining ground as we understand the urgency of seeing the interconnections between things and recognize that silo thinking reduces the capacity to address the significant complex threats we face. The primary threats to the stability and well-being of cities are climate change; the food, health, resource, poverty, and inequality crises leading to a security problem; and the financial crisis which limits the resources available to deal with them. Add to these a growing population that exerts pressure on everything and the mass movement of people across the globe, which can cause the identities of cities to shift, often with explosive results as frequently people living side by side have fundamentally differing views about how life should be lived. The inability to grasp the complexity of these risks and how to deal with them causes an intense governance and management problem. Finally, there is urgency and limited time to act, which is a crisis in itself. This is the risk landscape cities find themselves in; they encapsulate our vulnerabilities. It is an interlocking interdependent chain – a risk nexus. Understanding what this risk nexus is and how to deal with it is one of the cultural projects of our time.

2. The shared commons

Internationally, there is a demand for a reinvigorated public and shared commons. This is a social ethos that argues against our increasingly tribal and self-centred public culture. Among other things, it fosters spaces and places from parks to libraries that are free, noncommercial, and public. Places underpinned by this ethos can help retrofit conviviality and the habits of solidarity, and thus help to nurture our capacity to bond and to build social capital. Crucially, there is a contrast: with financial capital, the more you spend, the less you have; but with social capital, the more you spend on building social capital, the returns become ever greater. In time, the urban civility this fosters encourages individual and collective gestures of generosity. In turn, this self-generating process can create a virtuous cycle.

3. *Eco-consciousness*

All cities talk of sustainability. Every vision statement mentions combating the effects of climate change. Taking a helicopter view of cities worldwide, we see there are many good initiatives. Yet few cities make the hard planning choices to counteract the economic dynamics, spatial configurations, and physical forms that continue to make cities unsustainable in every sense. Cities have not been sufficiently imaginative in helping to change behaviour patterns, nor have they developed a new environmental aesthetic that inspires people to think afresh. Equally, 360-degree thinking has not embedded itself into decision-making circles so that it becomes a new common sense. As a consequence, regulations and incentive regimes are not clever enough to drive change. The necessary and dramatic retrofitting process still has a very long way to go, even though there are vast economic opportunities from being part of the fourth "lean, clean, green" industrial revolution. "Cradle to cradle" decision making remains far off.

4. *Healthy urban planning*

Urban planning that helps make you healthy by just navigating the city in day-to-day ways has not imbued planning disciplines. The cities we have built and continue to create make us unhealthy.

We now know about unhealthy urban planning: rigid "land use zoning" separates functions and prevents mixing uses, such as blending living, working, retail, and fun; "comprehensive development" that implements initiatives in one big hit often loses out on providing fine grain, diversity, and variety; "economies of scale" approaches tend to think that only the big is efficient or to produce things as if they were on a conveyor belt; and, lastly, the "inevitability of the car" can lead us to plan as if the car were king and people a mere nuisance. Walkable cities give you time and space to experience the city in visceral ways as part of being healthy and provide sensory satisfaction.

A healthy place is one where people feel an emotional, psychological, mental, physical, and aesthetic sense of well-being – where doing things that make you healthy happen as a matter of course and incidentally and not because you make a big effort. A healthy place throws generosity of spirit back at you. This makes you feel open and trusting. It encourages us to communicate across divides of wealth, class, and ethnicity. It makes for conviviality. And having trust is the precondition for learning, creativity, and innovation.

5. *Cultural literacy and intercultural thinking*

All our bigger cities are becoming much more diverse in their makeup. Multiculturalism as a planning concept and a policy is the predominant approach that acknowledges these differences. It highlights the need to cater to diverse needs. Interculturalism goes one step further and has different aims and priorities. It asks, "When we are sharing a city, what can we do together across our cultural differences?" It recognises difference yet seeks out similarities. It highlights that in reality most of us, when we look deep, are hybrids, and so it downplays ideas of purity. It stresses that there is one single and diverse public sphere, and it resources the places where cultures meet. It focuses less on resourcing projects and institutions that can act as gatekeepers and instead encourages bridge builders. In so doing, it does not consider that there is a cosy togetherness. It acknowledges conflicts and tries to embrace, manage, and negotiate a way through them based on an agreed set of guidelines of how to live together in our diversity and difference. In sum, it goes beyond a notion of equal opportunities and respect for existing cultural differences in order to achieve the pluralist transformation of public space, institutions, and our civic culture.

6. *Inequality and inclusivity*

Unequal societies create tension and resentment and lead to unfulfilled potential. A society of haves and have-nots does not harness the collective imagination and intelligence of its citizens or capture their energy and aspirations sufficiently. The trickle-down effect, it is increasingly recognized, is ineffective, and the gap between the rich and poor is growing. From the OECD to the annual Davos Summit, there are grave warnings of the looming effects and impacts of severe income disparity. It is seen as the most dangerous risk to social stability and well-being. It is corrosive, divisive, inefficient, and ethically unsound. Some degree of inequality may be good for an economy, creating incentives to work hard and take risks, but a high concentration of income gains among the most affluent is damaging. The Occupy protests demonstrated the increasing public anger that it has gone too far. It can be addressed, but only if we bend the market mechanism to public-good objectives.

7. *Intergenerational equity*

The *demographic time bomb* hangs over everything cities do. There will be pressure to isolate the aging population into retirement zones with housing adapted to their needs. More innovative places will seek to think through city making from an intergenerational perspective and develop adaptable housing forms that can be transformed through the lifecycle.

8. *The aesthetic imperative*

The city is a 360-degree immersive experience, and it communicates through every fibre of its being, its built structures, its natural forms, its activities, and its overall atmosphere. Its aesthetics engender an emotional response with psychological impacts. We can argue about ugliness and beauty, and crucially we should. This heightens our awareness of our surroundings and, in time, creates standards that, of course, are renegotiated. Yet there is usually more alignment on what works and what does not aesthetically.

The aesthetic imperative reminds us that every physical structure has an aesthetic responsibility to the environment and to the people in which it sits. People remember the pinpricks of ugliness spilling out from horrible buildings, misplaced urban design, or insensitive infrastructures throughout their lives. These have a negative impact leading to depression and other diseases, as work in environmental psychology shows again and again.

9. *Creative city making*

Creative city making seeks to address the escalating crisis cities face that cannot be solved by a business-as-usual approach, including the challenge of living together with great diversity, addressing the sustainability agenda, and addressing how cities can rethink their role and purpose in a changing world to survive well economically, culturally, and socially and to manage increasing complexity. It argues that curiosity, imagination, and creativity are the preconditions for inventions and innovations to develop as well as to solve intractable urban problems and to create interesting opportunities. Unleashing the creativity of citizens, organizations, and the city is an empowering process. Creativity harnesses potential, it searches out what is distinctive and special about a place, and it is a vital resource. It is a new form of capital and a currency in its own right. Creativity has broad-based implications and applications in all spheres of life. It is not only the domain of artists, those working in the creative economy, or scientists, though they are important. It also includes social innovators, interesting bureaucrats, or anyone who can solve problems in unusual ways. Cities need to create the conditions for people to think, plan, and act with imagination.

To make this happen requires a different conceptual framework. The capacity of a place is shaped by its history, its culture, its physical setting, and its overall operating conditions. This determines its character and "mindset." For too long there has been an "urban engineering paradigm" of city development focused on hardware. "Creative city making," in contrast, emphasizes how we need to understand the hardware and software simultaneously. In turn, this affects the "orgware" of a city – in other words, how to manage the city under these new conditions. Today, the essential element of the personality of many cities is their "culture of engineering." The attributes associated with this mindset are both positive and negative. It is logical, rational, and technologically adept; it learns by doing; and it tends to advance step by step and through trial and error. It is hardware focused. It gets things done. However, there is a weakness in this mindset in that it can become narrow, unimaginative, and inflexible and forget the software aspect, which is concerned with how a place feels and its capacity to foster interactions and to develop and harness skill and talent.

10. *An invigorated democracy*

In recent years, most things have been reinvented, such as how we do business, how we build cities, or how we entertain ourselves, and, of course, technology has moved apace in gigantic leaps enabling us to connect across the world in completely unforeseen ways. Yet our forms of representative democracy have remained largely the same for hundreds of years. Essentially, we vote for politicians to speak on our behalf every four years with little involvement in between, even though substantial efforts are made to consult citizens on local plans or, in some countries, to hold referenda on major issues. Clearly, this is not enough, as low participation in voting shows. Cities need to explore new ways of communicating with citizens so engagement with the civic can be reignited and policies can be cocreated. Here, the open data movement is one important initiative in making hitherto hidden information freely available, as are new ways of decision making such as citizen juries or other forms of participative democracy from online voting to town hall meetings. Overall, the key themes highlighted here are caring for others, celebrating and fostering distinctiveness and identity, and being open-minded in order to find solutions to any urban challenge.

Urbanity and its past

Urbanity and to be urbane has a combined economic, social, political, and cultural history which is useful to retrace and to recapture its best features for today. It represents an urban culture. The tradition of urbanity is essentially European, reflecting an attitude that emerged in the late Middle Ages in Italian city-states and in northern Europe and especially the Hanseatic League cities. It was led by merchants who tried to escape from the shackles and constrictions of feudalism to perform their trades in a less impeded way. In so doing, they became a vigorous group with their own political, economic, and cultural interests that successfully competed with the existing medieval order. They developed what became the bourgeois style of life, including their own learning and cultural institutions and norms and values. They were anti-feudal and, in their context, democratic. They were open and cosmopolitan and proud of their cities, and they invested in them. They reflected a new emerging economy based on trade and new methods of production, replete with new professional bodies, education and science institutions, and a focus on rational calculation. This gave citizens a sense of collective identity and shared solidarity reflecting an attitude to life. The city became more important than familial ties, clan bonds, or ethnicity. This allowed for greater mobility. This was a completely different worldview (Zijderveld 1998).

We are at a similar junction today, and a new knowledge-intensive economy is emerging. For our purposes here, the most significant points from this past urbanity are the commitment to the city rather than particular interests, a concern with its identity, and the openness. This represents a civic culture.

It is unwise to idealize this original bourgeois urbanity since it became more degraded, superficial, and consumption-oriented. Today, we sometimes interpret urbanity or associate it as a synonym for being suave, refined, or well mannered. Others see it as something to do with café culture, being somewhat cool, or a place with many cultural choices. Others of a more postmodern bent think that whatever a city is represents its urbanity. So if a place is a concrete jungle and dreadful, so be it.

Barriers to urbanity

Urbanity here is not merely a descriptive term but a programme for action. Today the world is more mobile, we identify with various places, and cities increasingly focus on attracting this workforce. These itinerant citizens have a different relationship to their cities. It is less intense or deep or long-term, and there is less commitment to place. Equally, the city usually has less power over key issues that determine its fate, such as education, transport, social welfare, or its ability to create its own rules, such as creating its own citizenship with appropriate rights.

At the same time, many independent voluntary and community structures, which were historically vital as the nervous system and mediating institutions of a city, are today relatively weaker as they are more reliant on national governments for survival. This makes our urban culture a reduced one because it has fewer levers to help develop citizens and so the civic. This decline in engagement is visible everywhere. It is reflected in low voting rates or the decline in trust, and so the invisible threads of connection – that make community work – weaken. Not surprisingly, therefore, when we think of urban culture, we think merely about the atmosphere, events, and arts of a city.

My notion of "civic urbanity" has more lofty aims. Yet how, in this overall context, do you develop a "civic urbanity" where place or our sense of anchoring is not what it used to be, or where virtual and real worlds blend more readily and where globalization changes the social life of communities in often negative ways so that they feel fragmented?

Being civic

Being civic is to be full citizens, which is people engaged with their cities in multiple ways on an ongoing basis in order to improve their lives and those of others. It is about feeling that *you* and *the city* merge into one as if it were part of you, as is every brick or blade of grass. The city owns you, and you own the city. Small day-to-day things, like the regular breakfast at a local café or the local dentist that you have seen for years, and the occasional larger events weave a web that, over time, feels like community. This familiarity happens imperceptibly and step by step as associations with place and people build up. These create memories, meanings, and histories. This identification takes time. It is why people often like places that to others are faceless, ugly, or soulless, because they can draw in many experiences. The bench where you had your first kiss, for instance, is why so much of peoples' identity is invested and embodied in them. There is a danger that this can entrap you and become claustrophobic as it closes you in, especially if the city in question is static and unchanging.

The young and, especially, the ambitious prefer to escape and may prefer a place that is on the move. This signals excitement, stimulation, and being where the action is. Yet acting in a civic way can, in principle, both deepen identity and develop and change the city, so making it feel alive and alert. The focus can be vast, from shifting the city to be green, to fostering local entrepreneurship, to getting different groups to mix or celebrate. Being civic throughout history has been linked to the democratic impulse. This implies being active and fosters a realm of debate and public discussion. Citizens at their best are makers, shapers, and cocreators of their evolving cities. They are producers of their places rather than merely consumers. The danger for most cities that need to attract the semi-permanents and itinerants with talent is that these individuals have little time to build commitment, direct involvement, participation, and loyalty. Instead, the buzz and liveliness is created for them, thus reinforcing the consumption bias.

To be civic often involves challenging the status quo and official institutions and being an activist. This builds up a civic society as a collection of engaged individuals often acting voluntarily, or as organizations and institutions that work together in ways that official bodies cannot or will not.

Generating civic urbanity

In closing, here are some guidelines to build civic urbanity:

- The first step is to bring the concept of civic urbanity into wider circulation to discuss its merits and possibilities.
- Spell out its potentially positive impacts to solve problems across a number of domains. This process builds evidence by showing examples of good practice.
- Persuade a city to explore civic urbanity in detail and to make this a policy programme. This will involve bringing a cross-departmental group together from physical planning, health, social affairs, economic development, environment, and culture.
- Develop a professional development programme to assess the city through various lenses, such as healthy urban planning or interculturalism.
- Undertake practical projects that embody the spirit of civic urbanity.

Note

1 For a review of these arguments, see Ash and Thrift (2007).

Reference

Ash, A., and Thrift, N. (2007) Cultural-economy and cities. *Progress in Human Geography*, 31(2), 143–161.

Zijderveld, A. C. (1998) *A Theory of Urbanity: The Economic and Civic Culture of Cities.* New Brunswick, NJ: Transaction Publishers.

3 We, European cities and towns

The role of culture for the evolving European model of urban sustainability

Svetlana Hristova

> We have been entrusted with something that is essentially fragile and perishable: life, the planet or the city. For the city is perishable. Its survival depends on us. . . . No institutional system survives without being sustained by a will to live together. . . .
>
> — Paul Ricœur (1991, 2)

In a world of increasing fragility and perishability, described by Ricœur, there is no doubt at least in one thing: that the old institutions of (pre)modernity need special attention to be sustained. The city is such an institution, deeply rooted back in the Euro-Asian past. In the framework of the concept of sustainable urban development, it is worthwhile to ask ourselves, is there something specific and unique that distinguishes European cities in their search for a new path to a sustainable future, something that can be described as an emerging model of a sustainable European city, *Europolis*? Even more specifically, what is the position of the small and medium-sized European cities (SMCs) in this process? And how can we understand their gradual reorientation over the last twenty years towards culture and creativity as the least exhaustible among all resources of urban development? Furthermore, does it mean that the "sustainable city" and the "creative city" coincide in the European context? Finally, to what degree can innovative thinking reenergize a city with the means of culture without diluting its authenticity? To examine these issues, this chapter presents the theoretical debates based on public documents and scientific contributions around the triple node of sustainable development, culture, and small and medium-sized cities.

Sustainable development

In the last three decades, globally raising social insecurities and ecological threats have brought about the issue of sustainable development as a universal concern, called by some the "Green Revolution."[1] As a response to these *trends that the planet and its people cannot long bear* (WCED 1987), two discursive streamlines have been developing, which mutually reinforce each other. On one hand, social and political scientists have been increasingly dealing with various aspects of endangered human life, culminating in the theory of the *risk society* that began to produce globally greater risks than wealth (Beck 1992). On the other hand, national governments have been coping with these problems on a practical daily basis with only sparse results. The crossing point of these simultaneous discourses and practices was the report *Our Common Future*, produced by the World Commission on Environment and Development (WCED), commissioned by the United Nations Organization, and issued in 1987. It is popularly known as the Brundtland Report after the name of the commission chairperson, Gro Harlem Brundtland, and has become a cornerstone in the contemporary conceptualization of sustainable development. Its introductory statement has been echoed repeatedly in almost any analysis dealing with this subject and has acquired the taken-for-granted status of an indisputable definition turned into a mantra: "Sustainable development . . . meets the needs of the present without compromising the ability of future generations to meet their own needs." However, taken out from its context, this statement does not say much. It may be interpreted in two ways (at least): as a need to save resources for the next generations, and as recognition of the present's limited possibilities to solve the sustainability problems that will be left to our successors. To solve this ambiguity, it is worthwhile to recontextualize the vision of sustainable development offered by the report.

While considering a number of specific threats to the planet, the report recognized explicitly the direct connection between economic development, environmental degradation, and social inequality. From that moment on, it would be impossible to think about solving any of these problems separately, as they all "are becoming ever more interwoven locally, regionally, nationally, and globally into a seamless net of causes and effects" (WCED 1987, 13, ¶ 15). This *new holistic developmental model*, denying the narrow preoccupations and compartmentalization of national economies, marked the beginning of a new stage of development – one of integrated global action and globally shared responsibilities. There are three important aspects of this model: *the imperative of limits*, *changed developmental aim*, and *differentiated approaches* to achieve these ends.

The developmental framework based on the *imperative of limits* includes both the limitations of human (technological and social) capacities and the limited capacity of nature.[2] While recognizing the global dangers and limitations for human existence and declaring many of the "present efforts to guard and maintain human progress, to meet human needs, and to realize human ambitions . . . simply unsustainable – in both the rich and poor nations" (WCED 1987, introduction, ¶ 25), the report declared openly the *rapid economic growth as a cause for the environmental damage*[3] and questioned both the purely profit-driven *goals* and the methods based on the uncontrolled use of nonrenewable resources. For this reason, the report advocated not just for accepting limits in all production, reproduction, and consumption patterns but for doing this in a *differentiated manner:* the developed and developing countries are expected to play different parts in the common scenario for sustainable development. To sum up, the report suggested – in the name of our common future – global redistribution of the causes, consequences, benefits, and responsibilities of development.

Finally, our sustainable future can be guaranteed only by a new development drive, the necessity to satisfy *human needs and aspirations*, declared to be *the major objective of development* (chapter 2, ¶ 4). These encompass both the needs of the present people and those of future generations. In the context of these new priorities, the well-known definition of sustainable development reveals its long-term importance: "In essence, sustainable development is a process of change in which the exploitation of resources, the direction of investments, the orientation of technological development; and institutional change are all in harmony and enhance both current and future potential to meet human needs and aspirations" (chapter 1, ¶ 15).

Sustainable urban development and the role of culture

In this new developmental model, cities hold a special place. Back in 1987, the report envisaged something that is a fact today: as of 2010, more than half of the global population lives in cities as a result of mass urbanization, which is taking place mainly in the developing world. This means that cities became central arenas for tackling problems of sustainable development, but in a differentiated manner given the significant differences between cities of the industrial countries and those of the developing world.[4] *Our Common Future* offers three far-reaching visions concerning the urban developmental path. First, there is the suggestion of *resetting the direction of urbanization*, by "taking the pressure off the largest urban centres and building up smaller towns and cities, more closely integrating them with their rural hinterlands" (WCED 1987, introduction, ¶ 73). To achieve this, the document foresees "examining and changing other policies . . . that work against the goals of settlements strategies" (introduction, ¶ 73). Second, the vision of *good city management* is based on the idea of *decentralization* of "funds, political power, and personnel to local authorities, which are best placed to appreciate and manage local needs" (introduction, ¶ 74). Third, the report contains the insight of *participative urban cultures* which will constitute new forms of urban governance: "[T]he sustainable development of cities will depend on closer work with the majorities of urban poor who are the true city builders, tapping the skills, energies, and resources of neighbourhood groups and those in the 'informal sector' " (introduction, ¶ 74).[5]

Finally, was culture totally excluded from this visionary report about our sustainable future? Culture, understood as a regulatory mechanism based on human values and motivations, is used implicitly in relation to human resource development: "to strengthen social, cultural, and economic motivations" as well as "to create new values to help individuals and nations cope with rapidly changing social, environmental, and development realities" (WCED 1987, introduction, ¶¶ 44, 45). Culture is also present in the analysis of human needs which are "socially and culturally determined and sustainable development requires the promotion of values that encourage consumption standards that are within the bounds of the ecological possible and to which all can reasonably aspire" (chapter 2, ¶ 5). Finally, although culture is not especially accentuated, its role is crucial as a new value promoter and pattern maker: "To successfully advance in solving global problems, we need to develop new methods of thinking, to elaborate new moral and value criteria, and, no doubt, new patterns of behaviour" (chapter 1, ¶ 45). However, in its narrow meaning (arts and creative expression), culture is absent in the report, which, more importantly, marks the cultural turn to a new developmental path.

Culture as both *overarching and underpinning human activities* appears explicitly in more recent thinking about sustainability, as exemplified by Jon Hawkes's metaphor of *culture as the fourth pillar of sustainability* (2001). This metaphor – compartmentalizing culture alongside economy, ecology, and social equity – has penetrated so deeply into current sustainability discourse that it blurs the authentic idea of Hawkes pleading for an integrated approach to culture as a factor for sustainable development. For him, it is not just an "arts-plus" matter but "a shared expression of, and commitment to 'a sense of meaning and purpose'" (Hawkes 2001, 13), and, for this reason, culture should be taken as an "aspect of the entire planning process that establishes the values upon which all planning is based" (36).

The ever-growing influence of culture for sustainable (urban) development is realized at different levels of European urban policies and practices, which can be called "European urbactivism." It uses creative expression and artistic activities for social purposes, defined as "artivism" (see Dragićević Šešić, Brkić, and Matejić, in this volume), based on new "green" values, combining bottom-up and top-down approaches, and supported by European funding instruments such as URBAN, URBACT, LEADER, LEADER+,[6] JESSICA and JEREMIE, and other programmes at regional and national levels, including independent initiatives of groups of towns and cities such as CENTROPE and the EuRegio Salzburg – Berchtesgadener Land – Traunstein (ESPON 2013).

The European urbactivism in the new urban age

> The issue of sustainable development is a fundamental theme of our civilization and a dominant topic on today's political agendas. Europe's major cities are both the nodal centres of economic growth and the living places for the most environmentally conscious segment of population.
>
> — Manuel Castells (1993, 253)

When during the 1990s visionary scientists outlined the growing potential of cities as a driving force for social change in *a new European society* (Castells 1993, 250) and saw *Europe as a community of cities* rather than a community of nations (Simioforides 1998, 144), these assertions were not very popular. European cities have been also viewed as core sites of growing pressure of competition scaled down from the nations to the level of cities and regions.[7] A decade later, after the 2005 and 2007 EU enlargements, these new visions were permitted into the official political discourse, highlighting the role of cities as generators of sustainable development and autonomous key actors at the European political stage.[8]

The first public document, an important benchmark of the increasingly proactive role of the European cities and towns in their search for sustainability, is the *Aalborg Charter of European Cities and Towns Towards Sustainability* (1994), adopted seven years later after the Brundtland Report as a direct response to the appeal of *Agenda 21* for support from local authorities. However, the *Aalborg Charter* moves beyond these documents, proposing a more contextual, European-specific model of urban sustainability.

In doing so, the *Aalborg Charter* develops further an *integrated and holistic* understanding of urban sustainability by adding a *European dimension* to it. Broadly defining sustainability as "the process of changing lifestyles, production, consumption and spatial patterns," led by the long-term goal "to base our standard of living on the carrying capacity of nature," the charter also adheres to the principle of *Our Common Future's differentiated approaches*, but applied at the level of single cities and towns.[9] This is one of the *Aalborg Charter*'s new insights: the recognition of *the city as the proper entity* to tackle the problems of sustainable development, being "both the largest unit capable of initially addressing the many . . . imbalances damaging our modern world and the smallest scale at which problems can be meaningfully resolved in an integrated, holistic and sustainable fashion" (¶ I.3).

Notably, culture is invisibly implicated in the concept of sustainability through the idea of lifestyles and human well-being, and, most of all, as "a creative, local, balance-seeking process extending into all areas of local decision-making." However, the potential of this fertile implication was not explicitly elaborated in the *Aalborg Charter,* which did not recognize it as a generator of sustainability. This oversight is noticeable even in more recent times. For example, in studies of the quality of life of European cities, culture is either missing or often degraded to entertainment, tourism, and restaurants (see, for example, EEA 2009; EC 2013). Nevertheless, the *Aalborg Charter* is a benchmark of the process of the emerging of *new spatial identities in Europe:* "We, European cities and towns . . . guardians of culture, heritage and tradition" (¶ I.1). The over three thousand signatures of various cultural origins and administrative types show the domination of SMCs: local authorities and municipalities, boroughs and counties, mountain communities, provinces and regions forming a *new imaginary community* of European sustainable cities and towns, extending beyond the EU and Europe itself.

The following decade demonstrated evidence of consistent political efforts at different levels (local, regional, national, interregional, and European) to embed and institutionalize the sustainability agenda by creating documental framework and funding mechanisms, providing feedback into the process with regular research and analysis, and offering qualitative objectives and indicators to assist cities in their self-monitoring. Cornerstones in this trend are the *Leipzig Charter on Sustainable European Cities*, adopted during the German EU Presidency in 2007 in support of the EU Sustainable Development Strategy and building on the Lille Action Programme (2000), the Rotterdam Urban Acquis (2004), and the Bristol Accord (2005); followed by the Marseille Statement (2008), the

Toledo Declaration (2010), and the "Reference Framework for European Sustainable Cities" under French leadership by the end of 2011 (see Duxbury, in this volume).

In 2012, the German Ministry of Transport, Building and Urban Development commissioned a study of the impact of the *Leipzig Charter* on integrated urban (district) development.[10] The research report *5 Years after the Leipzig Charter* (German Institute of Urban Affairs 2012) distinguished three major groups of countries based on their developmental strategies regarding deprived neighbourhoods:

Group A: countries with comprehensive national programmes for integrated urban development;[11]
Group B: countries with national or regional programmes or with national guidelines for integrated urban development;[12] and
Group C: countries with predominantly local approaches to integrated urban development.[13]

In comparison with the results of similar research carried out prior to the *Leipzig Charter – The Integrated Urban Development as a Prerequisite for Successful Urban Sustainability* (2007) – a very strong dynamic is observed in the adoption of national and local strategies towards sustainable development in the EU and beyond. All this can be summed up as a general trend of European *urbactivism*, which aims to produce *an ideal model for the European Sustainable City* and to translate it into practice,[14] characterized by the following traits:

(a) cities and towns reemerge as autonomous key actors, self-aware of their *strength, knowledge, and creative potential* to manage themselves *towards sustainability* (*Aalborg Charter*);
(b) they apply efficiency-oriented political actions using multiple tools based on expertise and analysis;
(c) they enact participatory process involving various sectors of the community;
(d) they acknowledge the priority of local sustainability-oriented solutions over universal one-size-fits-all formulas;
(e) they recognize the long-term prospective of sustainability as a process.

Generally, this is a continuous multilevel bottom-up and top-down process comprising cities and regions with different natural, economic, historical, and cultural assets in which gradually a new European society emerges – of shared responsibilities for our common future.

Small and medium-sized cities (SMCs) in the European spatial system

> Then I will proceed with my history, telling the story as I go along of small cities no less than of great. For most those which were great once are small today; and those which used to be small were great in my own time. Knowing, therefore,

that human prosperity never abides long in the same place, I shall pay attention to both alike.

— Herodotus, *The History*, I.vi

Since the times of Herodotus, the size of cities has mattered: big and great are thought to be the same in the changeable history of cities. But with the enlargement of the EU, the European urban realities are becoming less definable. For example, how small should the smallest towns in Europe be,[15] and how big the medium-sized cities[16] before turning into "big" ones? Comparisons, based on national definitions of cities, are hopelessly distorted across countries by differences in methodology (Dijkstra and Poelman 2012, 3). And which urban areas should be referenced: the administrative, the morphological, or the functional that usually do not coincide? With the development of more refined statistical methodologies, based on new technologies for monitoring and measuring territory, and the rapidly accumulated knowledge about it, the task of classifying European cities became even more complicated. Population size can be interpreted meaningfully only in regional and national contexts and, more specifically, from the viewpoint of area density, proximity to metropolitan centres, the functional profile of the settlements, and even historic criteria like city rights received through a charter.

Given the variety of research methodologies to identify and classify SMCs, different data are produced, but according to all of them SMCs prevail in the European urban landscape. According to the European Environment Agency (2006), approximately 40 percent of the EU citizens live in small urban areas (10,000 to 50,000 inhabitants), and 20 percent live in medium-sized cities (50,000 to 250,000 inhabitants). SMCs predominate in the European spatial system: there are nearly 1,000 urban centres with above 50,000 inhabitants and about 5,000 towns that have between 5,000 and 50,000 inhabitants (EEA 2006). In comparison, according to the Association of City Mayors, in Europe (including Russia, Ukraine, and Belorussia) there are only 264 big cities with more than 250,000 inhabitants. If we raise the upper threshold of European middle-sized cities to 500,000 inhabitants[17] (Giffinger et al. 2007), then 120 million or 40 percent of all urban citizens in Europe live in such settlements: "These 600 cities figure as the engines of economic development in space. Because of their large number they are the most decisive actors in order to make Europe more competitive and at the same time to make spatial development more sustainable" (19). The number of SMCs across the twenty-seven member states of the European Union is estimated to be 8,000 (ESPON 2013). In short, the SMCs are the basic and most numerous actors on the European urban stage, whose viability is a prerequisite for the sustainable development of European societies. Increasingly, they are gaining visibility and recognition in European regional policies and receive support from various DGs and initiatives of the European Commission.[18] In their efforts to find their own paths to a sustainable future, SMCs are rediscovering the infinite resources of their heritage, culture, and creativity.

As creativity and innovation emerge more frequently at the margins and boundaries rather than at the centres of systems (Isar and Anheier 2010), this is another strong argument in favour of the cultural re-imagination of European SMCs, often derailed in the margins of historical development.

Small and medium-sized cities: cultural potentials and challenges

The very foundations of European cities have been laid on *culture* – the Greek polis was born around ancient cults and temples as places that were necessary to express the connection and affection (*philia*) between citizens, as François de Polignac revealed in his research on the origins of the ancient Greek city (Polignac 1996). Today, however, SMCs are analyzed in a contradictory manner. On one hand, the specific conditions related to their size are seen as potential strengths of being more compact, less polluted, more equal, more transportable, and more manageable than the big ones due to their participative spirit, which stems from the openness, curiosity, and dialogical thinking of direct democracy. On the other hand, they are criticized as a wasteland of economic, social, and demographic stagnation, at least partially connected to their cultural conservatism and traditionalism: "Smallness is in the urban habitus; it is about ways of acting, self-image, the sedimented structures of feeling, sense of place and aspiration," and sometimes even of small-mindedness and parochialism (Bell and Jayne 2006, 5).

During the past decade's deepening structural crisis, there has been a growing interest in smaller cities as spaces for alternative economic revival, with the means of culture and creativity as the main resources of the competitive capacities of the SMCs: "Apart from city size, competitive urban development depends on the city's historical pathways and recent ability to elaborate a pro-active and effective urban development strategy fostering its innovative or creative power in cultural, social and economic terms" (Giffinger et al. 2007, 4). Culture and heritage, tourism and leisure – altogether, the experience economy – are seen as a pathway to sustainable urban life (Lorentzen and van Heur 2012). In a similar vein, Paul Knox and Heike Mayer (2009) explore their potential through a holistic approach into which the economy is interwoven innovatively with cultural traditions and local identities, ecological and social concerns.

When considering the potential of the European SMCs, we should remember that they once were epitomized by Ebenezer Howard's Garden City, an ideal of small healthy towns surrounded by agricultural belts and connected by efficient public transport, which could combine the positive characteristics of country and metropolis (better working opportunities, sociability, and cultural diversity of urban life). Visionary scholars such as Peter Hall and Colin Ward passionately supported the Garden City concept for its enormous potential to be an antidote to urban sprawl (Hall and Ward 2014).

Howard's ideas receive empowerment in the new information age within which the competitiveness of cities is determined by their ability to combine three key values: informational capacity, quality of life, and national and international connectivity (Castells 1993, 254). Equipped with the means of new information technologies and modern transport infrastructure, SMCs will better display their real potential for culture-led sustainable development.

Conclusion: imagining Europolis

> European cities are indeed cities and not just locales.
>
> — Manuel Castells (1993, 256)

Nowadays we observe so wide-spread practices of cultural regeneration of dilapidated urban areas that "it's hard to think of a city in Europe today which hasn't made some attempt to explore the possibilities, even if the latest initiatives are more dependent on private developers than the public sector" (Christiansen 2014). European cities have a long history back to ancient times, from where the legacy of urban public spaces and the traditions of public arts and agonistic spirit exemplified in the Olympic Games originate. And although the squares and city walls from the Middle Ages, which formed part of Max Weber's argumentation of the "ideal type" of European city, have changed their functions today, they can still present uniting symbolic elements in the spatial system of a shared European past.

However, for some decades culture as a factor for the sustainable development of European cities was not duly recognized in scientific discourses, policies, and practices. In this process of difficult recognition, the concept of sustainability has evolved along with an understanding of the various resources to be sustained, among which the least exhaustible are the cultural resources rooted in creativity. However, creativity is not an unquestionable concept. As revealed by numerous critiques of the guru of the "Creative Class" Richard Florida, creativity that is confined within the range of highly profitable activities reserved for the rich and well-off social groups tends to only intensify social and cultural exclusion. Obviously, such a creative city is not sustainable, as it is not able to "sustain life for those who live in it" (Derksen 2010) without prioritizing some of the citizens and ghettoizing others. An alternative model, stemming from European context, is expressed in Charles Landry's ideas of the creative city. Among the seven key factors of urban creativity – such as personal qualities, will and leadership, human diversity and access to varied talent, local identity, network dynamics, and urban spaces and facilities – Landry also noted organizational culture. It is exactly the culture that "represents the soil from within which creativity emerges and grows, and therefore provides momentum for development" (Landry 2000, 173).

Based on examples from Nordic countries, the 2011 INTELI report recognizes that "Florida's creative class approach is not a relevant framework for understanding the . . . regional growth in small city regions" and that there is "a need to reconcile the analysis of the 'business-climate' and 'people climate' approaches, but essentially to recast the creative capital model in line with the specific characteristics of small territories, privileging alternative concepts and indicators, such as quality of life and natural, historical-cultural and symbolic amenities" (INTELI 2011, 47).

This organizational culture makes cities different in their regional, national, and international contexts. As long as cultures are different in their organizational capacity to produce and sustain innovative thinking, the main question remains open: if creativity can be taught, can it be learned? Some researchers even confirm that cultures within the European Union can be distinguished into two major groups: recipients of innovation and producers of innovation. Therefore, the existing tension between the organizational field of sustainability, insisting on the preservation and reproduction of existing cultural codes, and innovation, aiming at their change and upgrading, cannot be easily overcome or ignored (see Johannisson, in this volume). As Yudhishthir Raj Isar and Helmut K. Anheier elaborate, "alongside and perhaps gradually superseding this image of the individual genius kissed by the muse," there is a new idea of creativity: "it is based on the understanding that the likelihoods of creativity to emerge, of creative acts to be recognized, and of both leading to innovation, yielding sustained change, are all closely linked to the organization of economy and society as well as patterns in the cultural domain itself" (Isar and Anheier 2010, 4).

The idea of Europolis gives us such hope, based equally on the continuing European urban traditions and the capacity for their innovation. Europolis – the metaphoric name given once to the Romanian city Sulina in the mouth of the Danube, designed as a European project at the end of the nineteenth and beginning of the twentieth century[19] – is a failure now, when the economic and political interests have flown away, together with the waters in the artificial channel. But the idea of Europolis – as an epitome of the potential of the European compact, human-sized cities to achieve a balanced urban growth addressing human needs, and based on the caring capacity of nature – is still alive. It is a city in process, a sustainable city in development, a learning city capitalizing upon the European culture as a fundamental asset. As Manuel Castells notes, "For that reason they should be able to manage the articulation between the space of flows and space of places, between function and experience, and between power and culture. In this way they could recreate the city of the future by building on the foundations of their past" (1993, 256). Looking *back in the future*, as Castells suggests, Europolis gives us a promise for more sustainable development.

Notes

1 See, for example, Peach (2013).
2 As ¶ 27 in the introduction specifies, these are "not absolute limits but limitations imposed by the present state of technology and social organization on environmental resources and by the ability of the biosphere to absorb the effects of human activities" (WCED 1987). See also chapter 2.
3 "Environmental concern arose from damage caused by the rapid economic growth" (WCED 1987, ¶ 34); "both the rich and poor nations . . . draw too heavily, too quickly, on already overdrawn environmental resource accounts. . . . They may show profit on the balance sheets of our generation, but our children will inherit the losses. We borrow environmental capital from future generations with no intention or prospect of repaying" (¶ 25).
4 The report depicts various problems of cities in industrial countries (deteriorating infrastructure, environmental degradation, inner-city decay, and neighbourhood collapse), which are conceived as an issue "ultimately . . . of political and social choice," and recognizes that developing countries "are not in the same situation. They have a major urban crisis on their hands" (WCED 1987, introduction, ¶ 72).

5 For a wider discussion and recommendations, see chapter 9 of the WCED (1987) report.
6 Although these two initiatives have been focused on the rural development of fifteen EU member states until 2006 and the enhancement of urban-rural links, they also involve SMCs (ESPON 2013).
7 A good review of the existing scientific discussion on different aspects of the competition between cities is presented in Giffinger et al. (2007).
8 Yet this is only a relative trend as long as cities are not directly part of transnational governance structures and have no seats or votes at bodies such as the European Parliament, Council of Europe, and the like (see Isar, Hoelscher, and Anheier 2012).
9 "As each city is different, we have to find our individual ways towards sustainability. We shall integrate the principles of sustainability in all our policies and make the respective strengths of our cities and towns the basis of locally appropriate strategies" (*Aalborg Charter 1994*).
10 It was carried out in the twenty-seven member states of the EU; the six candidate countries at that time (Iceland, Croatia, Macedonia, Montenegro, Serbia, and Turkey); and the two European Free Trade Association (EFTA) states, Norway and Switzerland.
11 These are Belgium, Denmark, Finland, France (case study Lyon – Pentes de la Croix-Rousse), Germany, Ireland, Italy, the Netherlands (case study Amsterdam – Nieuw-West), Sweden, Switzerland, and the United Kingdom.
12 They include Austria, Bulgaria, Cyprus, Czech Republic, Hungary, Iceland, Latvia, Lithuania, Malta, Montenegro, Norway, Poland, Portugal, Romania (case study Timişoara), Slovenia, and Spain.
13 Croatia, Estonia, Greece, Luxembourg, Macedonia, Serbia, Slovakia, and Turkey fall in this group.
14 See www.rfsc-community.eu/about-rfsc/background/.
15 The bottom line varies between two thousand and five thousand inhabitants in different European countries.
16 The thresholds coming from different national contexts, commonsense observations, and research methodologies vary between 250,000 and 500,000 inhabitants.
17 This coincides with the threshold, accepted in the assessment of the global megatrends of urban worlds, according to which "smaller cities" are those that count less than 500,000 inhabitants (EEA 2014).
18 These are such projects and programmes as "Cities of Tomorrow" and "CityStars" (DG Regional Policy), "Smart Cities and Communities Initiative" (DG Energy), and "European Green Capital" (DG Environment). The European Observation Network for Territorial Development and Cohesion (ESPON), founded by the European Commission in 2007, is also an important mechanism for development of well-informed territorial cohesion policies, focused on the SMCs, among others (ESPON 2013).
19 The history of this city is presented in the documentary film *Europolis* (2010), by the Bulgarian producer Kosta Bonev. Originally, *Europolis* was the title of a novel (1933) written by the Romanian writer Eugeniu Botez under the artistic pseudonym Jean Bart, predicting the advancement and decay of the port city Sulina.

References

Aalborg Charter (1994) Available at: www.sustainablecities.eu/fileadmin/content/JOIN/Aalborg_Charter_english_1_.pdf [Accessed 25 January 2014].

Beck, U. (1992) [1986] *Risk Society: Towards a New Modernity*. New Delhi: Sage. Translated from German *Risikogesellschaft: Auf dem Weg in eine andere Moderne*, 1986.

Bell, D., and Jayne, M. (eds.) (2006) *Small Cities: Urban Experience Beyond the Metropolis*. London: Routledge.

Bundesministerium für Verkehr, Bau und Stadtentwicklung (BMVBS) and Bundesamt für Bauwesen und Raumordnung (BBR) (eds.) (2007) *Integrated Urban Development – a Prerequisite for Urban Sustainability in Europe*. Berlin and Bonn: BMVBS and BBR.

Castells, M. (1993) European cities, the informational society, and the global economy. *Journal of Economic and Social Geography*, 84(4), 247–257.

Christiansen, R. (2014, December 16) Are cultural districts transforming the world? *The Telegraph*. Available at: www.telegraph.co.uk/culture/art/architecture/10997543/Are-cultural-districts-transforming-the-world.html?mobile=basic [Accessed 25 September 2014].

Derksen, J. (2010) How high is the city, how deep is our love. *Fillip*,12. Available at: http://fillip.ca/content/how-high-is-the-city-how-deep-is-our-love [Accessed 16 December 2014].

Dijkstra, L., and Poelman, H. (2012) Cities in Europe: The new OECD-EC definition. *Regional Focus* (January). Brussels: European Commission.

European Commission (EC) (2013) The RFSC Conference 2013: Towards a European Model of a Sustainable City, 9 October 2013, Brussels (Belgium). Available at: www.rfsc-community.eu/events/rfsc-conference-2013/ [Accessed 15 May 2014].

European Environment Agency (EEA) (2006) *Urban Sprawl in Europe: The Ignored Challenge*. Report no. 10. Luxembourg: Office for Official Publications of the European Communities. Available at: www.eea.europa.eu/publications/eea_report_2006_10 [Accessed 18 March 2013].

European Environment Agency (EEA) (2009) *Ensuring Quality of Life in Europe's Cities and Towns: Tackling the Environmental Challenges Driven by European and Global Change*. Luxembourg: Office for Official Publications of the European Communities. Available at: www.eea.europa.eu/publications/quality-of-life-in-Europes-cities-and-towns [Accessed 18 October 2014].

European Environment Agency (EEA) (2014) *Assessment of Global Megatrends – an Update. Global Megatrend 2: Towards a More Urban World*. Copenhagen: EEA. Available at: www.eea.europa.eu/publications/global-megatrends-update-2 [Accessed 18 October 2014].

European Observation Network, Territorial Development and Cohesion (ESPON) (2013) *TOWN: Small and Medium Sized Towns in Their Functional Territorial Context*. Applied Research 2013/1/23, Inception Report, Version 04/07/2012. Luxembourg: ESPON and KU Leuven. Available at: www.espon.eu/export/sites/default/Documents/Projects/AppliedResearch/TOWN/TOWN_Inception_report_July_2012.pdf [Accessed 18 October 2014].

German Institute of Urban Affairs (ed.) (2012) *5 Years after the LEIPZIG CHARTER: Integrated Urban Development as a Prerequisite of Sustainable City: Integrated Urban Development in the EU Member States and Its Candidate Countries*. Available at: www.hallbarastader.gov.se/Bazment/Alias/Files/?5yearsliepziegcharter [Accessed 5 August 2014].

Giffinger, R., Fertner, C., Kramar, H., Kalasek, R., Pichler-Milanović, N., and Meijers, E. (2007) *Smart Cities: Ranking of European Medium-Sized Cities*. Vienna: Centre of Regional Science, Vienna University of Technology. Available at: www.smart-cities.eu/download/smart_cities_final_report.pdf [Accessed 18 June 2014].

Hall, P., and Ward, C. (2014) *Sociable Cities: The 21st-Century Reinvention of the Garden City*. London: Routledge.

Hawkes, J. (2001) *The Fourth Pillar of Sustainability: Culture's Essential Role in Public Planning*. Melbourne: Common Ground Publishing and Cultural Development Network.

INTELI (2011) *Creative-Based Strategies for Small and Middle-Sized Cities: Guidelines for Local Authorities*. Lisbon: INTELI. Available at: http://urbact.eu/fileadmin/Projects/Creative_Clusters/documents_media/URBACTCreativeClusters_TAP_INTELI_Final.pdf [Accessed 2 February 2012].

Isar, Y.R., and Anheier, H. K. (2010) Introduction. In H. K. Anheier, Y.R. Isar, and C. Waterman (eds.), *Cultural Expression, Creativity and Innovation* (pp. 1–16). The Cultures and Globalization Series, no. 3. London: Sage.

Isar, Y.R., Hoelscher, M., and Anheier, H. K. (2012) Introduction. In H. K. Anheier, Y.R. Isar, and M. Hoelscher (eds.), *Cities, Cultural Poliy and Governance* (pp. 1–12). The Cultures and Globalization Series, no. 5. London: Sage.

Knox, P., and Mayer, H. (2009) *Small Town Sustainability: Economic, Social, and Environmental Innovation.* Basel: Birkhauser Verlag AG.

Landry, C. (2000) *Creative City*. London: Comedia.

Lorentzen, A., and van Heur, B. (eds.) (2012) *Cultural Political Economy of Small Cities.* Abingdon: Routledge.

Peach, J. (2013) Infographic: Cities embracing the green revolution. *ThisBigCity.net*. Available at: http://thisbigcity.net/infographic-cities-embracing-the-green-revolution/?goback=%2Egde_69691_member_251273463 [Accessed 16 December 2014].

Polignac, F. de (1996) *La Naissance de la cité grecque: Cultes, espaces et société, VIIIe–VIIe siècle av. J.-C.* Paris: Editions La Decouverte.

Ricœur, P. (1991, October 29) Un entretien avec Paul Ricœur: « La Cité est fondamentalement périssable: Sa survie dépend de nous » [interview with R. P. Droit]. *Monde*, 48(14541), 2.

Simioforides, Y. (1998) The Greek city and the new urban conditions. English summary of the Italian article: La citta Greca e le nuove condizioni urbane. In F. Infussi and Y. Simioforides (eds.), *Programma Heracles: Le sfide della citta greca* (pp. 141–154). Athens: Linora.

World Commission on Environment and Development (WCED) (1987) *Our Common Future*. Report of the Brundtland Commission. Oxford: Oxford University Press.

4 Culture, quality of life, and sustainable urban development

Aldo Milohnić

In times of economic crisis, the requirements imposed upon cultural workers and civil servants in the field of culture are so dramatic that they call for a serious assessment of the situation and careful planning of a strategy to produce convincing arguments defending components of the welfare state (including culture). In principle, investments bring with them new employment opportunities to all sectors of society, and culture is not an exception. On the other hand, cultural policy should not indulge in uncritical praise of the "industrialization" of culture, given that many researchers have already pointed out its negative effects – for example, a specific kind of qualitative "collateral damage" caused by the cultural, creative, and entertainment industries that, in turn, affects the nature of work in culture. Furthermore, as pointed out by Lorentzen and van Heur (2012), "creative work," an integral part of cultural and creative industries, can be seen as a euphemistic discursive curtain veiling less "fashionable" facts behind the spectacle: "Highlighting the entrepreneurial dynamics of creative work, the policy discourse on the creative industries tends to emphasize the innovative potential of these sectors, but mostly downplays the reality of underpaid labor and the recurring exclusions along the lines of gender, ethnicity, age and class within the creative industries" (Lorentzen and van Heur 2012, 2). In other words, culture is not only the number of tickets sold; it is in actuality more about creativity, in its intellectual – and not commercial – dimension. In essence, culture's impact resides in its ability to challenge the norm as well as its ability to contribute to the well-being of national and local communities.

This chapter begins by discussing controversial interactions of cultural and economic values, as an example of so-called city regeneration projects. In the second part, cultural and creative industries, very popular notions in mainstream cultural policy and research literature, will be confronted with a more universal concept of "the right to the city." After this theoretical contextualization, the chapter discusses two outstanding Slovenian examples of urban community gardening – an important aspect of the "right to green," a concept which has gained a lot of interest in contemporary theories of so-called hands-on urbanism.

From "city regeneration" to the "right to the city" model

Concepts of cultural and creative industries, when they are applied to urban planning, should be deployed carefully and with a "healthy dose" of distance – as can be traced in a number of city regeneration projects. A typical problem of many urban regeneration and redevelopment projects, based on the economic potential of cultural/creative industries, is an uneven social and spatial development of cities. For instance, there is a vast body of scientific literature on both the advantages and the disadvantages of regeneration projects in cities on the Iberian Peninsula in the 1990s and early 2000s (Barcelona, Bilbao, Valencia, etc.).

Statistically speaking, the new cultural policies of Spanish cities were successful in attracting millions of tourist visits, generating thousands of new employment positions, improving the communication infrastructure, and so forth. In the case of Barcelona, approximately five years after the implementation of the "Strategic Plan for the Cultural Sector," an interim evaluation (*Balanç* 2003) listed many positive economic results of the plan. However, as of 2006, local decision makers became aware that the impressive statistical data also had an adverse effect – which has been pointed out by many independent researchers (Jordi Borja, Joan Roca, Josep Lluis Montaner, Manuel Delgado, and Eduard Bru, among others) and civil society organizations (such as Forum Ribera del Besós, Federació de Associacions de Veins, and others, including critical voices from important cultural institutions such as the Museum of Contemporary Art of Barcelona [MACBA], Fundació Tapies, and the Centre of Contemporary Culture of Barcelona [CCCB]). According to these researchers and organizations, the instrumentalization of culture in order to achieve political, social, and economic goals brought many harmful side effects, not only for those who were active in the field of culture (and did not participate in the mainstream cultural production) but also for the people of Barcelona in general – for instance, conversion of the city into a "theme park" for tourists and consumers diminished the quality of life of local inhabitants; massive speculation with land and housing contributed to increased stratification of the urban population; accelerated gentrification and privatization of public spaces; and devastated cultural heritage and traditional working-class neighbourhoods.[1]

The city of Bilbao, sometimes called the "Basques cultural miracle," is another instructive example of a cultural and urban development model. Bilbao gradually developed a new cultural image for the city in the 1990s, most importantly through its investment in the Guggenheim Museum. Local researchers think that the "Guggenheim effect" was so strong that the mere opening of the museum in 1997 brought the city about a million extra visitors per year and thousands of new jobs in the fields of culture, tourism, and other service activities (Plaza 2000; Haarlich and Plaza 2010). Many critical voices, however, see the cultural development model of Bilbao as an example of the economically motivated "culture of spectacle," "neoliberal urbanization," and "American cultural imperialism" (McNeill 2000; Swyngedouw, Moulaert, and Rodriguez 2002). In their opposing opinion, the main purpose of the museum should have been to serve the cultural needs of the local community rather than to generate only economic benefits (Rodriguez, Martinez, and Guenaga 2001). Furthermore, it generated significant social and spatial exclusion. In reference to Bilbao and other European cities, Swyngedouw, Moulaert, and Rodriguez (2002) allude to "urban islands," "fragmented city," and "suburbanization of poverty" (576). Nowadays, many of the Spanish flagship projects, especially big museum institutions (not only in Bilbao and Barcelona, but also in Valencia, León, Santiago de Compostela, and elsewhere), are suffering radical budget cuts as a consequence of austerity measures taken by national, regional, and local authorities, and they are facing a sharp drop in number of visitors (see Buck 2013).

The first precondition for the successful utilization of cultural policies based on the fast development of cultural and creative economies should be a strictly individual and site-specific approach that is sensitive to historical, social, economic, demographic, ecological, and other important aspects of the city in question. Furthermore, urban cultural policies should not only be context-specific; they also must be sensitive to the size of a city. Small and medium-sized cities have their own advantages, which often provide unique opportunities for cultural development. These advantages of small and medium-sized cities are usually not the same as the characteristics of metropolis-like cities, which are typically used as examples in mainstream academic research in urban cultural policy development: "Urban qualities of small cities include not only lower prices of land and housing, but also lower degrees of pollution, congestion and crime. Walkability and natural surroundings allow for lifestyles that are different from those in most metropolises. Smaller cities, in that respect, are often seen as highly appealing to families, certain types of artists, and elderly people" (Lorentzen and van Heur 2012, 6).

Instead of a "city regeneration" approach, which can be useful in cultural policy-led urban development of cities only to a certain extent, contemporary cities might turn to a more universal concept of "the right to the city," as introduced by French philosopher Henri Lefebvre in 1960s and rediscovered in 1990s by David Harvey, nowadays one of the leading scholars in urban and cultural geography. In Lefebvre's vision, "the *right to the city* . . . can only be formulated as a transformed and renewed *right to urban life*" (Lefebvre 2000, 158). As Lefebvre described eloquently in his book *Le Droit à la ville*, modern cities should not be just places for endless accumulation of capital, an urban habitat of people subordinated to daily reproduction which can end up only in reification of their life. People can emancipate themselves only if they are ready to take their destiny in their own hands. In other words, the modern city will be the true *home* of all its inhabitants providing that it is their own daily creation. As we can see, according to the Lefebvrean idea of the right to the city, notions like *creation* or *creativity* are not just empty phrases (as we can trace, for instance, in many policy papers or political speeches); they are embedded in the very essence of the modern urban life of the emancipated inhabitant of the city. This idea was productively re-actualized by David Harvey, whose treatise against neoliberal appropriation of urban life (and at the same time a *plaidoyer* for the right to the city as a political slogan) calls for an urban resistance to that tendency. His explanation of the concept is that

> [t]he right to the city is far more than the individual liberty to access urban resources: it is a right to change ourselves by changing the city. It is, moreover, a common rather than an individual right since this transformation inevitably depends upon the exercise of a collective power to reshape the processes of urbanization.
>
> (Harvey 2008, 23)

There is an ongoing discussion, both in academic and in activist circles, of an *authentic* reading of the concept of the right to the city, on its supposed *trivialization*, on a more *radical* or rather *reformist* interpretation, and so forth. No matter in which direction this debate will continue, it is already clear that Lefebvre-Harvey's concept has found its followers and admirers not only in Europe but also in Asia, Africa, Latin America, and elsewhere (Parnell and Pieterse 2010; Faetanini 2011; Gorgens and van Donk 2011; Brenner, Marcuse, and Mayer 2012). It indicates per se that the concept has a lot of potential. It is socially inclusive and opens up new perspectives on urban development. And last but not least, it enables a bottom-up approach to interventions in the urban daily life of contemporary cities.

Urban community gardening as "the right to green"

From my perspective, urban community gardening, as a concept and a practice, represents an important aspect of the right to the city model. Historically, it started to grow in the nineteenth century with the rise of industrialization and increasing proletarianization of the rural population coming to big cities in search for work. Workers in expanding urban and industrial centres practiced urban gardening as a source of self-produced food, especially in English industrial cities, but also in German, Austrian, and other European urban centres (Krasny 2012). A long tradition of urban gardening can also be traced in the United States and elsewhere. In the twentieth century, especially in times of war atrocities and economic crises (such as the Great Depression in 1930s), urban gardening was an important self-help strategy of maintaining subsistence for workers' families in modern cities. A very instructive example of how an unplanned and abrupt scarcity in raw materials and commodities can radically change national agriculture is the so-called *Período Especial* (Special Period) in Cuba. As in many other countries, Cuban agriculture was highly dependent on mechanization and chemicals. After the dissolution of the Soviet Union at the beginning of the 1990s, Cuban exports (to the former Eastern Bloc) dropped by 80 percent, and the country faced a terrible shortage in its fuel supply as well as previously imported fertilizers and pesticides. All these unfavourable conditions created a fertile ground – metaphorically and literally – for urban agriculture. Instead of using chemicals in agricultural production and transportation of food from one region to another, Cubans started to produce fruits and vegetables close by: in small gardens in the city (for individual consumption) and in large gardens in suburbs (for trade on green markets). They rediscovered old knowledge and skills concerning organic, chemical-free food production (Bohn and Viljoen 2012). As nicely shown in Ray Burley's 2006 film, *Cuba: The Accidental Revolution – Part 1: Sustainable Agriculture*, urban gardens were created everywhere: in vacant lots, school playgrounds, backyards, and so on. Thanks to that radical turn, Cuba developed a respectable programme in sustainable agriculture and managed to recover its food production almost entirely by the end of the 1990s.

With the present economic crisis (which is at the same time a social and environmental one), the self-production of food and sustainable development have become a substantial challenge in our highly urbanized societies. Crisis is not only a negative social and economic phenomena; it is also an opportunity to rethink and to reform existing city policies on urban gardening. Nowadays, urban gardening is a common worldwide practice: from Europe to the Americas, from Berlin and London to Seoul and Hong Kong.

Urban gardening in Slovenia, as part of a growing process of industrialization and urbanization of the country, started at the beginning of the twentieth century. After WWII, urban gardening in Slovenia was dominated by leisure and recreation activities, while vegetable gardening was rather neglected. A new wave of urban gardening occurred in the 1980s, and then it slowly declined in the following decades. In the last few years, it is again becoming popular, and there are many bottom-up initiatives promoting and practicing urban agricultural projects. In the latter part of this chapter, two case studies of urban community gardening in the two biggest Slovenian cities are briefly presented: *Beyond a Construction Site* in Ljubljana and *Urban Furrows* in Maribor. Both examples are vivid, ongoing, participative projects that bring together local communities, artists, a broad spectrum of experts, members of different minorities and marginalized social groups, activists, and so forth. It is interesting that the general frame of both projects is culture: *Beyond a Construction Site* started in August 2010 as part of a cultural festival in Ljubljana, while the *Urban Furrows* project was launched in the frame of the Maribor European Capital of Culture in 2012. *Beyond a Construction Site* is an example of urban community gardening (originally an initiative of the Obrat Cultural Association) and only one in a series of projects experimenting with temporary use of (usually degraded) urban areas and/or objects in Ljubljana. The *Urban Furrows* project is a social and cultural experiment exploring possible modes of cohabitation of various ecological initiatives (i.e. community gardening, a seed library, sustainable local supply, and promotion of cycling), minority empowerment programmes (i.e. Roma, homeless people), and alternative production (i.e. the association Centre for Alternative and Autonomous Production [CAAP] as a platform for various social economy initiatives).

Beyond a construction site

Beyond a Construction Site is a grassroots initiative in urban community gardening of the Obrat Cultural Association in Ljubljana.[2] It is an excellent example of the temporary use of space for that purpose – an abandoned construction lot in the middle of a residential neighbourhood in the centre of Ljubljana. The construction site was closed at least six years before the initiative started. The owner of the plot is the Municipality of Ljubljana. The urban community garden was launched in August 2010 in the frame of the programme "Garden by the Way," part of the *Mladi levi* (Young Lions) festival. The cultural festival was an appropriate context to start up the project, but, from the very beginning, Obrat planned for a longer duration, not only the two weeks of the festival. The festival supported the initiative in the second year of its existence as well. Furthermore, having the festival as a "back-up" of the initiative, local authorities were more likely to give the organization permission for a free-of-charge use of the land. Now the project has a temporary agreement with the municipality, which is prolonged every year until the anticipated start of construction.

The initial idea of *Beyond a Construction Site* was the creation of a "green area," not necessarily urban community gardening. In the course of time, however, it went in that direction, following the ideas and expectations articulated in discussions with inhabitants of the neighbourhood. Launching the project within the framework of a cultural event offered an excellent opportunity for promotion of the project. On the other hand, people from Obrat were aware of a potential danger that a grassroots project can become an ingredient of an urban regeneration plan of developers, although that kind of commodification was not an original intention of the initiative.

There are certain rules in the community garden, but they are not rigid, and some of them can be consensually changed if the experience shows that certain corrections are necessary. There are, however, some basic rules all participants in the project have to obey. For instance, members of the group should construct and maintain their own vegetable bed or, if they are not able to do this on their own, at least should help in the construction of their bed. Furthermore, no chemicals are allowed for use in gardening. In addition to taking care of their own garden, participants also have to assume responsibility for taking joint care of the common area.

In this project, re-creation of social interactions is equally important as re-creation in terms of physical activities in a green area and gardening. It is also an educational playground for children from young families who otherwise could not learn the useful skills of planting vegetables and maintaining a garden and who would probably not connect with each other at all. As the participants in *Beyond a Construction Site* share all resources (space, tools, water), it is necessary to work with differences and contradictions. However, even occasional conflicts among participants might be productive if they are transparent and manageable: "In contrast to an idea of community based on a notion of identity and belonging ('being in'), which is necessarily always exclusionary for 'others', we are striving for a community that produces more open and fluid relationships that foster a sense of 'being with' " (Obrat 2012, 327).

Urban community gardening is not just a gardening activity, and it is not only about producing vegetables and fruits for the community's own needs; it is also about developing social interactions among people in a neighbourhood. People who do not know each other – who do not have any kind of social relations – come together, collaborate, and exchange knowledge, skills, views, and so forth. As Obrat (2012) notes, "And just as gardening is an activity, community is also an activity – a process and an effort to establish relationships and links, not only among the participants in *Beyond a Construction Site*, but also *beyond* the project itself" (328). The project is an opportunity for everyone involved to learn about urban gardening and ecology as well as about sharing the management of a space and various participatory processes. They organize regular meetings, host public lectures and discussions, publish bulletins, and search for connections with similar initiatives in Slovenia and abroad.

According to Urška Jurman and Polonca Lovšin from the Obrat Cultural Association, the temporary use of space is a big motivation for participants: they want to use both the land and the time allotted to them in the best possible way. For them, it is still better to have a community garden for a limited time than not to have it at all. They are adaptable and, if necessary, could move to another location. A bigger problem is that the local authorities in Ljubljana are not interested in increasing the number of locations in the city where urban community gardening could be planned as a permanent activity.

Many people are skeptical about the ecological aspect of the urban gardening; they ask themselves, "Is the food produced in urban gardens really healthy?" Of course, fruits and vegetables should not be consumed if the soil is contaminated by, for instance, long-time pollution from heavy metallurgy or the chemical industry. On the other hand, food and vegetables produced industrially are treated by many pesticides and fertilizers, some of them are genetically modified, they are often transported from a very long distance, and so forth. In the case of the Ljubljana and Maribor community gardens, specialized institutes examined samples of soil, and findings confirmed that in both gardens the soil is absolutely appropriate for producing healthy food.

It is important that the garden is not far away from users' apartments. The urban way of life is very dynamic, and many people have no time to travel almost every day out of the city to maintain their gardens. Furthermore, for the concept of urban community gardening, it is important that social interactions developed through common gardening are transferred to the neighbourhood as a whole. In other words, people who maintain a community garden together are more likely also to spend time together outside of the garden.

This small urban community garden in the centre of Ljubljana might be an important case study also for possible reform of the city's policy of organizing and leasing small garden plots.[3] The city's current policy in that field represents a rigid, standardized, and too formalistic approach:

> The criteria for leasing a garden plot (age and income) are turning the gardening areas into social ghettos; the gardens are being relegated to the city outskirts; and moreover, it is predicted that there will be a drastic reduction in the amount of land allotted to community gardens.
>
> (Obrat 2012, 328)

Furthermore, a decline in the amount of square kilometres of gardening land in Ljubljana is an obvious negative trend: 2.67 km^2 in the mid-1990s, 1.86 km^2 in 2005, and only 1.3 km^2 in 2008 (Jamnik, Smrekar, and Vrščaj 2009, 64–65).[4]

Urban Furrows

The *Urban Furrows* project is a social and cultural experiment exploring possible modes of cohabitation of various ecological initiatives, minority empowerment programmes, and alternative production.[5] Conceptual preparation of the project started in fall 2010, and in the beginning of 2011 the head of the project (Marta Gregorčič) and her colleagues proposed *Urban Furrows* to the organizational team of the Maribor European Capital of Culture 2012. It emerged from the theory of social ecology (Murray Bookchin, Daniel Chodorkoff) and an understanding of culture in the broadest possible way, including the culture of consumption, culture of self-subsistence, and culture of solidarity. "Empowerment of people," says Chodorkoff, "is the real goal of any authentic process of development. Social ecology calls for the primacy of these socio-cultural criteria over the economic" (Chodorkoff 2005, 37).

For social ecology, the notions of society and ecology are interdependent: there is no ecology without its social dimension, and vice versa. All activities within the project *Urban Furrows* are interconnected. The urban community garden is only one segment within a much broader programme that includes small educational-recreational gardens in schools and homes for the elderly, a seed library, and a sustainable local food supply programme. Other programmes are Teleport, a cyclist and pedestrian friendly community; the Rhizome Collective, concerned with modern national and international migration policies; Ethnomobile, which is researching Roma and ethnic issues; First Step, addressing homelessness and the state of discarded youth; and the Centre for Alternative and Autonomous Production (CAAP). This chapter focuses only on the urban community gardening part of the *Urban Furrows* project.

In Slovenia, there is a very strong tradition of self-produced fruits and vegetables – towns are rather small and populated mainly by people of different generations coming from rural areas. On one hand, this situation has a great potential for urban gardening, and, on the other hand, horticulture is an important part of Slovenian culture in a broader sense. The project of the urban community garden in Maribor found its historical roots in a settlement of railway workers on the right-hand bank of the Drava River. The colony was established in 1863, and it was populated by workers from the refurbishing and repair workshops of the Austrian Southern Railway, operating services from Vienna to Trieste. Each building had up to eight apartments, and for each of them there was a small plot with fruit trees and a garden with vegetables. Workers' families living in the colony shared among themselves common garden sheds, tree lanes, the town wells, and so on. This colony from the mid-nineteenth century has been preserved to this day.

Maribor European Capital of Culture 2012 was an excellent context to develop such a complex ecological, social, and cultural project. Furthermore, this big cultural event was a chance to multiply the effect and the echoing of the *Urban Furrows* project, and it was much easier to get in touch with local inhabitants and decision makers within the local authorities. Because of this special situation, the association established on the initiative of the project team and local participants was able to obtain a large piece of land on an appropriate and desired location in the densely populated part of the city for a long time (at least ten years). It was also important that the project received funds from the budget of Maribor European Capital of Culture 2012 and that they were spent in a very rational manner. According to Matej Zonta, coordinator of the urban community garden within the *Urban Furrows* project, if we compare their expenses with the budget for a top-down creation of small, individual, rentable gardens in Ljubljana, they spent five times less per capita than the Municipality of Ljubljana.

There was a whole process of educating people about urban gardening and, at the same time, building up a new community by exercising mutual communication, searching for agreements, and enacting consensual decision making. "We always say: on our community garden, we don't cultivate only plants, but also (or even more) social relations," says Zonta.

The urban community gardening component of the project is a success story, at least for now.[6] There are many similar initiatives in Slovenia. They are eager to learn from the experiences of the urban community garden in Maribor, and they are willing to initiate this obviously good practice in their own cities and towns (Domžale, Kamnik, Kranj, Izola, Murska Sobota, Celje, Slovenj Gradec, etc.). However, they all have the same problem: with only a few exceptions, all they can get from their municipalities is a piece of land on an unsuitable location and no funds at all to establish the most basic infrastructure. Another effect of the prevailing bureaucratic logic of local authorities in Slovenia is a ghettoization of urban gardeners. A typical way of thinking goes like this: let's give elderly, poor, and other marginalized social groups a piece of rather useless land to have their "playing ground" somewhere in the suburbs, as far away from the city centre as possible. Furthermore, urban community gardening is in constant conflict with various financial interests: speculation with land, the profit interests of developers and investors in the construction sector, taxes and other contributions collected by the local authorities and national tax offices, and so on. But the present economic crisis has led to a big drop in the construction sector and has provided a great chance for urban community gardeners to establish their gardens closer to city centres.

On the national level, one can hear only promises that urban gardening will be supported in a systemic manner: "As for now we don't see any concrete actions of the Slovenian government in that direction," says Zonta. Nevertheless, the ecological movement, the economic crisis, and a lot of urban gardening initiatives appearing at the same time have contributed significantly to raising general awareness and support among the public (as consumers) and the authorities (as decision makers) for urban community gardening.

Conclusion

"Quality of life is difficult to quantify," says Daniel Chodorkoff. "But the goal of development must be focused on providing people with the security that their basic needs, like adequate food and shelter, will be met, as well as what are often intangible areas that are reflected in a sensibility. Well-being undoubtedly requires a degree of economic security, but it rests more on a sense of socio-cultural security" (Chodorkoff 2005, 36). It seems that projects of urban community gardening have a certain potential to foster a sustainable urban development of contemporary cities on the global scale. By proclaiming the "right to green," they are fighting for a higher level of the quality of urban life, for the "right to the city," and for the right to re-create urban agriculture as an integral part of the culture of modern cities.

Urban community gardening is participative by definition – a typical way of establishing a new project is that an initial group invites anyone who is interested (in particular, residents of the neighbourhood) to take an active part in the decision-making process about the development and organization of the space. They are also social laboratories: lessons learned from case studies examined show that there is no standardized norm or technique of participation in community-based projects. Due to different contexts and circumstances, methods of participation should be reinvented with every new project of urban community gardening. In other words, cultivating plants is at the same time cultivating social relations.

An important dimension of *Beyond a Construction Site* and *Urban Furrows* is the raising of awareness of city decision makers as well as local inhabitants about all the positive aspects of the urban community gardening – they are at the same time social and cultural experiments and examples of good practice. And they are important laboratories of social ecology, which will hopefully influence future urban development policies not only in Ljubljana and Maribor but also in all other Slovenian cities. If so, one hopes that the negative trend of the declining amount of gardening land in Ljubljana and other cities will be halted and returned to growth.

Notes

1 For more elaborated insights, see, among others, Balibrea 2005, 2007; Milohnić 2009.

2 This section is based on an in-depth interview with Urška Jurman and Polonca Lovšin, active members of the Obrat Cultural Association and the *Beyond a Construction Site* urban community garden. The author conducted the interview on 6 February 2014. Additional sources are Obrat's bulletins and a chapter in the book *Hands-on Urbanism* (Obrat 2012).

3 Members of the association are active in popularization of the community garden approach. Some of their most recent activities in that direction include workshops and other educational programmes for children (summer 2014) and an exhibition on hands-on urbanism and the right to green, followed by a public lecture by Elke Krasny (both in September 2014).

4 In another source (Potrebuješ 2013, 25) we can find different numbers: 2.58 km^2 in 1995, 2.12 km^2 in 2006, and 1.67 km^2 in 2010. Differences are probably caused by methodological reasons, but a negative trend is obvious in both cases.

5 This section is based on an in-depth interview with Matej Zonta, coordinator of the ecological part of the *Urban Furrows* project. The author conducted the interview on 7 February 2014.

6 The first phase in the lifetime of the project was recently documented in the film *Green Utopia*. A trailer of the film (with English subtitles) is available at: www.youtube.com/watch?v=WGq7k67L6F8 [Accessed 23 March 2014]. According to Zoran Zobovič, president of the association Urbani eko vrt (Urban Eco Garden), the urban community garden is still doing very well, and local inhabitants show a constant interest in joining the project. For the moment, 150 candidates for renting a small piece of land are on the association's waiting list (Zobovič 2014).

References

Balanç 1999–2003. (2003) Barcelona: Ajuntament de Barcelona, Institut de Cultura. Available at: www.bcn.es/plaestrategicdecultura/pdf/Balanc_ICUB_1999_2003.pdf [Accessed 23 March 2014].

Balibrea, M. P. (2005) Barcelona: Del modelo a la marca. In J. Carrillo and I. Estella Noriega (eds.), *Desacuerdos 3: Sobre Arte, Políticas y Esfera Pública en el Estado Español* (pp. 263–267). Barcelona: Arteleku, MACBA, Universidad Internacional de Andalucia.

Balibrea, M. P. (2007) The brand and the past: Strategies of the struggle for social space in postindustrial Barcelona. In YProductions (ed.), *Producta50* (pp. 23–24). Barcelona: Generalitat de Catalunya.

Bohn, K., and Viljoen, A. (2012) Laboratories for urban agriculture: Havana and Milwaukee. In E. Krasny (ed.), *Hands-on Urbanism 1850–2012: The Right to Green* (pp. 226–236). Hong Kong: MCCM Creations.

Brenner, N., Marcuse, P., and Mayer, M. (eds.) (2012) *Cities for People Not for Profit: Critical Urban Theory and the Right to the City*. New York: Routledge.

Buck, T. (2013) Spain's austerity programme thwarts regional cultural ambitions. *Financial Times*, October 13. Available at: www.ft.com/cms/s/0/bb91acce-31cb-11e3-817c-00144feab7de.html#axzz2ig5Xb5mn [Accessed 23 March 2014].

Burley, R. (2006) *Cuba: The Accidental Revolution – Part 1: Sustainable Agriculture* [film]. 45 min. Produced by "The Nature of Things" television programme, Canadian Broadcasting Corporation.

Chodorkoff, D. (2005) Redefiniranje razvoja [Redefining development]. *ČKZ*, 22(219), 34–43. English version available at: www.social-ecology.org.

Faetanini, M. (ed.) (2011) *Urban Policies and the Right to the City in India*. New Delhi: UNESCO and Centre de Sciences Humaines.

Gorgens, T., and van Donk, M. (2011) From basic needs towards socio-spatial transformation: Coming to grips with the 'Right to the City' for the urban poor in South Africa. Conference paper. Cape Town: Isandla Institute. Available at: http://isandla.org.za/publications/152/ [Accessed 23 March 2014].

Haarlich, S. N., and Plaza, B. (2010) Das Guggenheim Museum von Bilbao als Symbol für erfolgreichen Wandel – Legende und Wirklichkeit. In U. Altrock, S. Huning, T. Kuder, H. Nuissl, and D. Peters (eds.), *Symbolische Orte: Planerische (De-)Konstruktionen* (pp. 150–165). Berlin: Reihe Plannungsrundschau 19.

Harvey, D. (2008) The right to the city. *New Left Review*, 53(September–October), 23–40.

Jamnik, B., Smrekar, A., and Vrščaj, B. (2009) *Vrtičkarstvo v Ljubljani* [*Gardening in Ljubljana*]. Ljubljana: ZRC.

Krasny, E. (2012) Hands-on urbanism 1850–2012: The right to green. In E. Krasny (ed.), *Hands-on Urbanism 1850–2012: The Right to Green* (pp. 8–36). Hong Kong: MCCM Creations.

Lefebvre, H. (2000) The right to the city. In H. Lefebvre, *Writings on Cities* (pp. 147–159). Oxford and Malden: Blackwell Publishers.

Lorentzen, A., and van Heur, B. (2012) *Cultural Political Economy of Small Cities*. London: Routledge.

McNeill, D. (2000) McGuggenisation? National identity and globalisation in the Basque country. *Political Geography*, 19(4), 473–494.

Milohnić, A. (2009) Primerjava mest v širšem evropskem prostoru: Skandinavija (Helsinki), osrednja Evropa (Bern) in južna Evropa (Barcelona) [A comparison of cities in a broader European space: Scandinavia (Helsinki), mid-Europe (Bern) and south Europe (Barcelona)]. In A. Milohnić and M. Breznik, *Primerjalna analiza razvoja kulture v izbranih evropskih mestih* [*A Comparative Analysis of Development of Culture in Selected European Cities*]. Ljubljana: Peace Institute.

Obrat (2012) Beyond a construction site: A community-based garden intervention in a neglected urban space in Ljubljana. In E. Krasny (ed.), *Hands-on Urbanism 1850–2012: The Right to Green* (pp. 322–330). Hong Kong: MCCM Creations.

Parnell, S., and Pieterse, E. (2010) The 'right to the city': Institutional imperatives of a developmental state. *International Journal of Urban and Regional Research*, 34(1), 146–162.

Plaza, B. (2000) Evaluating the influence of a large cultural artefact in the attraction of tourism: The Guggenheim Museum Bilbao Case. *Urban Affairs Review*, 36(2), 264–274.

Potrebuješ, U. (2013) *Analiza in usmeritve za prostorsko organizacijo vrtičkov na območju Mestne občine Ljubljana* [*Analysis and directions for spatial organization of allotment gardening in the area of the Municipality of Ljubljana*]. Ljubljana: University of Ljubljana, Faculty of Civil and Geodetic Engineering.

Rodriguez, A., Martinez, E., and Guenaga, G. (2001) Uneven redevelopment, new urban policies, and socio-spatial fragmentation in metropolitan Bilbao. *European Urban and Regional Studies*, 8(2), 161–178.

Swyngedouw, E., Moulaert, F., and Rodriguez, A. (2002) Neoliberal urbanization in Europe: Large-scale urban development projects and the new urban policy. *Antipode*, 34(3), 542–577.

Zobovič, Z. (2014) Interview with Zoran Zobovič by Radio Slovenia, 8 June 2014. Available at: http://ava.rtvslo.si/predvajaj/urbani-eko-vrtovi-koscek-zemlje-za-mariborske-vrtickarje/ava2.174279529/ [Accessed 27 November 2014].

Section II

Europolis as a project

Envisioning more sustainable cities

5 European cities as cultural projects

Where is culture in urban sustainability policy?

Nancy Duxbury

> Cities are not just economic engines, they are unrivalled as providers of the basic ingredients for quality of life in all its senses: environmental, cultural and social.
>
> — European Commission, Directorate General for Regional Policy (2011, 42)

Cities are identified as key sites of action in global policy initiatives to recognize the important role of culture in sustainable development and to integrate culture in policy contexts at all levels (e.g. UNESCO 2013). Within Europe, culture has played a driving role in urban redevelopment, economic and branding strategies, and social inclusion initiatives. A variety of European-level policy statements cite the important cultural roles played by European cities, and there is a generalized refrain of "European cities as cultural projects" (Duxbury et al. 2012). But how strong is the link between culture and urban sustainability? To what extent are cultural dimensions included in policy/planning frameworks for "sustainable cities" in Europe? How do these policy frameworks suggest pathways for culture in building more sustainable urban futures?

Europe-wide frameworks concerning urban sustainable development have been informed by collective statements and initiatives from three main sources: *European cities*, primarily developed through meetings/conferences and advanced by associations of cities and the Council of Europe's Congress of Local and Regional Authorities; *national ministers of EU member states with responsibilities for urban development*, meeting under the auspices of the Council of Europe or the European Union; and the *European Commission*, through EU policy, reports, and funding programmes. From time to time, policy documents also cite European research programmes intended to inform policy and practice, and extra-European reference points such as charters and principles collectively developed by cities.

This chapter examines the major European urban sustainability planning/policy frameworks and guides (see Table 5.1), and it investigates to what extent and how cultural considerations are incorporated. The analysis focuses on key documents from the mid-1990s. This choice of time span means that greater emphasis is placed on the role of the European Commission and less on the Council of Europe, but it recognizes the Council's important role in the 1980s in Western Europe and in the 1990s in Eastern Europe and the enlarged EU.

Table 5.1 Chronology of key European policies, statements, and programmes

Year	*Local and regional authorities*	*National Ministers*	*European Commission*	
	European Cities: Sustainable Cities & Towns Campaign (1994–2013) / Movement of Sustainable Cities and Towns (2013–)	*European Union: Ministers Responsible for Urban Development*	*European Union: Policy Statements*	*European Union: Funding Programmes*
1994	*The Aalborg Charter* (27 May 1994)			URBAN (support for urban regeneration initiatives) (1994–1999)
1997		*Towards an Urban Agenda in the European Union*		
1999			*Sustainable urban development in the European Union: a framework for action* (COM(98) 605 final) adopted	
2000		Lille 2000 meeting		URBAN II (support for urban regeneration initiatives) (2000–2006)
2001			EU Sustainable Development Strategy (Göteborg Strategy)	
2002				URBACT (2002–2013)
2003				
2004	*The Aalborg Commitments ("Aalborg + 10")*	Rotterdam 2004 meeting		

2005		Bristol 2005 meeting / Bristol Accord	Review of the EU Sustainable Development Strategy
2006			
2007		*Leipzig Charter on Sustainable European Cities*	*EU Territorial Agenda*
2008		Marseille 2008 meeting – Decision to create a tool to translate into practice common sustainability goals and the Leipzig Charter objectives	
2010		Toledo 2010 meeting	
2011		EU Reference Framework for Sustainable Cities (RFSC) – toolkit (testing phase in 2011, finalized in Jan. 2013) and conference, "Towards a European Model of a Sustainable City" (Oct. 2013)	*Territorial Agenda of the European Union 2020*
2012			
2013	Movement of Sustainable Cities and Towns launched *European Sustainable Cities Platform*		Increasing emphasis on "place-basing of policy," coordination and integration of sectoral policies (Campos 2013)

Analysis was guided by two questions: how are cultural considerations incorporated (or not) in these documents? How does this text point to potential pathways for developing more explicit connections with culture in urban sustainability policy and initiatives? This chapter aims to help bridge a divide often observed at the local level: cities continue to struggle to understand how to integrate culture into urban sustainability, and municipal cultural administrations are still not regularly involved in large urban development processes and issues.

The Aalborg Process: the European cities movement

The trail of city-driven efforts to conceptualize urban sustainability and advance policy and planning practices begins with the *Aalborg Charter* (1994) and *Aalborg Commitments* (2004), which were developed and carried forward through the Sustainable Cities and Towns Campaign (1994–2013) and, since 2013, the Movement of Sustainable Cities and Towns. This city-driven movement is currently encapsulated in the European Sustainable Cities Platform (ESCP), an information portal that brings together "all relevant partners working on issues around sustainable cities" and provides a one-stop shop for local communities (ESCP 2014).

The Sustainable Cities and Towns Campaign was a bottom-up movement with a threefold mission: "to support the exchange of experience between cities, collect information on the activities undertaken at the local level and serve as interface between the European Union and the local sustainability movement" (all quotes from ESCP 2014). The Campaign played a key role in "defining what a sustainable European city should look like" (*Aalborg Charter* 1994) and in "setting out a process for making this vision a reality" (*Aalborg Commitments* 2004). Together, these two documents form a framework for movement toward sustainability in cities/towns, and they are generally referred to as the *Aalborg Process for Local Sustainability*. The European Union (2008b) credits these efforts for introducing the *sustainable city* concept into the European policy realm.

The *Aalborg Charter* was an urban *environmental* sustainability initiative, declaratory in nature, approved by participants at the first European Conference on Sustainable Cities and Towns in Aalborg, Denmark, on 27 May 1994 (signed by over three thousand local authorities from more than forty European countries, with Spain and Italy representing significant proportions of these signatories). The Charter was inspired by the Rio Earth Summit's Local Agenda 21 plan and was intended to contribute to the European Union's Environmental Action Programme, "Towards Sustainability."

In the *Aalborg Charter*, European cities and towns are described as "centres of social life, carriers of our economies, and *guardians of culture, heritage and tradition*" (1994, 1 [emphasis added]). Cities and towns are also viewed as "key players in the process of changing lifestyles, production, consumption and spatial patterns" (1) – dimensions closely linked to cultural expression, activities, and related dynamics. However, when the Charter goes on to note local authorities' intentions "to integrate people's basic social needs . . . [and] work towards improving the quality of citizens' lifestyles" (3), there is no explicit mention of culture as an aspect of these "basic social needs."

Ten years after the release of the Charter, the *Aalborg Commitments* were developed at the 4th European Conference on Sustainable Cities and Towns (Aalborg) to articulate "a common understanding of sustainability" and, consequently, "to develop a framework to be used at the local level that would better articulate how to embed sustainability across municipality sectors" (ESCP 2014). The Commitments comprise a list of fifty qualitative objectives organized into ten themes and represent a structured approach, requiring the signatory to comply with "time-bound milestones" (see ESCP 2014). The Commitments were positioned in reference to Local Agenda 21 and the forthcoming EU Thematic Strategy on the Urban Environment (2006), and have about seven hundred signatories, with local authorities from Spain and Italy dominant and including signatories from outside Europe (in Niger, Egypt, Tunisia, Morocco, and Senegal).

Within the *Aalborg Commitments*, culture is mentioned under two themes – (1) urban planning and design, in reference to urban cultural heritage, and (2) social equity and justice, in reference to equitable access to cultural activities:

- Planning and Design (no. 5) – "We are committed to a strategic role for urban planning and design in addressing environmental, social, economic, health and *cultural* issues for the benefit of all."
    - Sub-item no. 4: "ensure appropriate conservation, renovation and use/re-use of our *urban cultural heritage*."
- Social Equity and Justice (no. 9) – "We are committed to securing inclusive and supportive communities."
    - Sub-item no. 2: "ensure *equitable access* to public services, education, employment opportunities, training, information, and *cultural activities*."

(*Aalborg Commitments* 2004 [emphasis added])

In 2013, building on twenty years of the Campaign, and recognizing the array of initiatives now in play, the Movement of Sustainable Cities and Towns, also referred to as the "European Sustainable Cities Movement," was launched. This new framework is rooted in four principles of "Sustainability DNA":

1. Holistic thinking – In light of the many environments to take account of in a city ("global, local, *cultural*, urban, rural, political and social . . . [the] necessary ingredients of our holistic environment"), policy initiatives should span across multiple dimensions of the city with decisions made across sectors to further "the interests of the whole as defined at city level" and to "represent the city's vision in its entirety" rather than the advancement of one particular area;
2. Sustainability skills – The knowledge to make responsible choices and understand the consequences of choices for both current and future generations;
3. Partnerships – Among municipalities and across sectors; and

4 Research – Although focused largely on hard science and technology, the social sciences are seen to play an important role in providing "a better understanding of the social processes that cause people to change their behaviour for the good."

(ESCP 2014 [emphasis added])

In this framework, the cultural environment is explicitly viewed as one of the "necessary ingredients" within a holistically conceptualized city that should be considered when making policy and decisions.

The Congress of Local and Regional Authorities of the Council of Europe is an additional player among the "city" voices, acting as a political "bridge," encouraging cities to take advantage of European programmes and funding opportunities, and expressing cities' perspectives and priorities to the EU. This official political conduit complements the more operational focus of the Campaign/Movement and has been highly relevant in the past.

National ministers with responsibilities for urban development: the urban agenda

An explicit "European consensus" on principles of urban development – referred to as the *Acquis Urbain* – has emerged through an ongoing, intergovernmental process of more than two decades and the practical experiences gained through projects financed by the European Regional Development Fund (ERDF) programmes since 1989. The political trail features six informal ministerial meetings on urban development between 2000 and 2010, which have "shaped common European objectives and principles for urban development" and helped forge "a culture of cooperation on urban affairs" among member states, various European bodies, and "urban stakeholders" through European city organizations (EC-DGRP 2011, 7). The chief outcomes of this process have been the *Leipzig Charter on Sustainable European Cities* (2007), consequently linked to the objectives of Europe 2020 through the Toledo Declaration (EU 2010), and the operationalization of the *Leipzig Charter* through the development of the Reference Framework for European Sustainable Cities (2011–2013).[1]

The European Ministers responsible for urban development signed the *Leipzig Charter on Sustainable European Cities* on 24 May 2007. With this charter, the twenty-seven member states, for the first time, "outlined an ideal model for the European Sustainable City and laid the foundations for an integrated urban policy" (EC 2014). The *Leipzig Charter* stated that to achieve the objective of sustainable cities, an integrated approach to urban issues must be chosen, and that European structural funds should be made available for local projects embracing this integrated approach (EU 2007).

In the *Leipzig Charter*, the Ministers declare that

> all dimensions of sustainable development should be taken into account at the same time and with the same weight. These include economic prosperity,

> social balance and a healthy environment. At the same time attention should be paid to cultural and health aspects . . . [and to] institutional capacity in the Member States.
>
> (EU 2007, 1)

European cities are described as possessing "unique cultural and architectural qualities" and functioning as "centres of knowledge" but also suffering from an array of social and environmental problems. The Charter argues that to fulfill their functions as "engines of social progress and economic growth" (cf. Lisbon Strategy), the social balance within and among them must be maintained, cultural diversity must be ensured, and high quality in the fields of urban design, architecture, and environment must be established.

In the *Leipzig Charter*, the concept of the *Baukultur* of a city's living environment is introduced, understood in a broad sense as "the sum of all the cultural, economic, technological, social and ecological aspects influencing the quality and process of planning and construction" (EU 2007, 3). While the approach is viewed as "particularly important" in the preservation of architectural heritage such as historical buildings and public spaces, the Charter argues that this holistic approach should also be applied to the city as a whole and its surroundings. The *Baukultur* approach also stresses citizen inclusion in planning processes (EC-DGRP 2011).

At the 2010 informal meeting of Urban Development Ministers, held in Toledo, Spain, the topic of "integrated urban regeneration" was discussed further. The background reference document for the meeting, prepared by Spain, is structured using the "classical viewpoint of the multiple dimensions of sustainability (economic, social, environmental, cultural and governance)," indicating an elevation of the place of culture within sustainability. European cities and heritage are positioned as both *key elements* and *repositories* of "the rich and varied European history and culture" (EU-Spain 2010, 4). Within a section outlining "the key features of the integrated approach," the adoption of a holistic mode of thinking is primary. This involves considering the city as a whole, incorporating transversal or multidimensional approaches, and aligning different policy areas and resources across "all the multiple dimensions of sustainability – economic, social, *cultural* and environmental" (5 [emphasis added]).

However, despite prominently including culture in the sustainability framework, the report has difficulty going beyond a focus on physical heritage and public space rehabilitation. The report notes that preservation of "the historical and cultural heritage of the city, particularly its architectural heritage and the 'Baukultur' " is generally understood as necessary to keep alive "the collective memory that is characteristic of the European city model" (4). The "inhabitability and attractiveness" of this physical heritage is also necessary to "keep it really alive" (4). The report goes on to discuss building rehabilitation and improving degraded public spaces to increase their attractiveness and local attachment to the urban environment and community. This process of rehabilitation and physical improvement, it notes, will contribute to "cultural enrichment" and citizenship by fostering "the values of democracy, coexistence, exchange, civic progress, diversity, living together and freedom . . . key factors in the culture of the European city, which are expressed most effectively in the public realm" (4). No mention is made of cultural creation and expression or access to culture for citizens.

In 2008, the European Ministers decided to build "a reference framework for the sustainable city, in a spirit of solidarity, for the application of the Leipzig Charter" (EU 2008a, 5). This free tool would promote and "translate into practice the common sustainability goals and the Leipzig Charter objectives" (EC 2014) and give cities a tool to help them reach "a European vision of 'integrated urban development' " (personal communication, C. Guichard, Architecte Urbaniste de l'État, CEREMA, France, 18 March 2014).

Self-described as "a toolkit for the integrated approach," the Reference Framework for European Sustainable Cities (RFSC) is designed for local authorities and "intended to offer guidance and support for improving decision-making and action on sustainability" (EC 2014). Acknowledging that each city has its own history, cultural identity, economic background, and specific issues, the RFSC is not normative or prescriptive but offers selected questions and indicators to ensure that every political dimension, including cultural ones, had been taken into account (personal communication, C. Guichard, 18 March 2014).

The RFSC results from a collaboration between EU member states, European institutions, and European organizations representing cities and local governments. In the development process, workgroups studied more than seventy existing tools, methods, and reference frameworks from different cities, and twelve were assessed in depth (EC 2014). Projects identified as "good practices" were also examined. These analyses formed the basis for building the question grid and indicator list in the RFSC. The testing phase, with sixty cities, occurred in 2011, and the "final" version of the tool was released in January 2013.

The RFSC is built around twenty-five core objectives derived from four pillars of sustainability – economic, social, environmental, and governance (EC 2014) – and offers three main pathways of use:

1 *Develop your strategy/project* – The user picks from a list of "actions/objectives" those that best suit local priorities (additional actions/objectives can be created), indicates their level of commitment to each objective (on a scale of: Initiation – Commitment – Maturity), and then "checks the relevance" of the chosen objectives. If some domains of sustainability are not addressed, the system advises the user that the strategy "is likely to be unbalanced regarding urban sustainability." The inclusion of cultural objectives may lead some users to consider cultural objectives or dimensions that may not have been considered during initial planning phases.
2 *Check the integrated approach* – The user lists the "priorities" of an integrated planning initiative (using the same list of objectives) and ranks each by level of importance (Low – Medium – High). The system then provides a "results" analysis and plots the positioning of each of the selections on a circle graph with twenty-two axes, providing at a glance a visual representation of the

importance given to each selection, and a profile of the (un)balanced nature of the overall project (e.g. a deficiency of attention to cultural objectives and sub-objectives results in a "gap" in the graph).

3 *Monitor progress* – This section of the tool suggests an array of indicators, linked to each Core Objective and sub-objective, and the user chooses to select each or not (see Table 5.2). It is in this section that the conceptual ideas are translated into operational monitoring strategies.

The RFSC promises to shape the basis for sustainable development planning practice going forward, as a minimum of 5 percent of EU funding support is now earmarked for integrated sustainable urban development, with an implicit expectation that the RFSC will be used in this work (EC 2013).

Among the core objectives defining "a European vision of the sustainable city," culture is explicitly referenced in three ways in the RFSC – (1) a social quality (cultural diversity), (2) social activities, and (3) a dimension of the built environment:

Objective 12 is "Promote cultural and leisure opportunities and ensure access for everyone," with four sub-objectives:

- encourage and value cultural diversity;
- support and encourage cultural and artistic creation and exchange;
- ensure broad, affordable, and equal access to culture for everyone;
- provide leisure and sports facilities.

Objective 17 is "Preserve and promote the high quality and functionality of the built environment, public spaces and urban landscape," with two of four sub-objectives referencing heritage and architecture respectively:

- identify, preserve and promote the existing heritage according to the local and cultural context;
- promote and enhance the architectural quality of urban landscapes, public spaces, and the built environment.

Linked to the cultural objectives, the RFSC system provides an array of indicators to measure conditions and monitor change (see Table 5.2). The cultural indicators were inspired by a number of sources, including the Urban Audit data from Eurostat, the Global City Indicators Framework, and the UK Audit Commission (personal communication, C. Guichard, 18 March 2014). While it is not possible in this chapter to conduct a full critique, the two Core Objectives appear to be supported by their sub-objectives, but the alignment between the objectives and the indicators selected for monitoring them is problematic. Analysis of the suggested indicators shows a primary focus on physical assets and, due to the mix of culture with leisure and sport, very general relations between some of the indicators and "culture." No indicators explicitly measure the sustainability of local culture (although the combination of government and resident support,

Table 5.2 Cultural indicators within the Reference Framework for European Sustainable Cities (February 2014)

	Objectives[a]	*Indicators*	*Cultural dimension*[b]
Social Domain			
Core Objective	**Promote cultural and leisure opportunities and ensure access for everyone**	K 14 – Percentage of municipal budget allocated to cultural and sporting facilities	Built environment
Secondary Objectives	Support and encourage cultural and artistic creation and exchange	SI 186 – Percentage of jobs in the cultural sector	Social activity
		SI 187 – Number of persons employed in the culture and entertainment industry	Social activity
		SI 188 – Engagement in the arts	Social activity
		SI 189 – Culture and leisure expenditure	Social activity
		SI 190 – Civic participation in the local area	Social activity
	Ensure broad, affordable, and equal access to culture for everyone	SI 191 – Satisfaction with cultural facilities	Built environment
		SI 192 – Total loans of books and other media per resident	Social activity
		SI 193 – Use of public libraries	Social activity
		SI 194 – Citizens' satisfaction with the local community/ municipality with regard to the level of cultural, recreational, and leisure services	Social activity
	Provide leisure and sports facilities	SI 195 – Proportion of the area in sports and leisure use	Built environment
		SI 196 – Satisfaction with public parks and gardens (green spaces)	Built environment
	Take specific measures to support language skills for foreigners	*(None)*	Cultural diversity / social quality
	Promote multicultural exchange between the inhabitants (festivals, fairs, culture exchange, etc.)		Cultural diversity / social quality

Environment Domain			
Core Objective	**Preserve and promote the high quality and functionality of the built environment, public spaces, and urban landscape**	K 20 – Satisfaction with public spaces	Built environment
Secondary Objectives	Identify, preserve, and promote the existing heritage according to the local and cultural context	SI 269 – Number of listed buildings	Built environment
		SI 270 – Share of listed buildings restored	Built environment
		SI 271 – Expenditure on listed buildings	Built environment
	Promote and enhance the architectural quality of urban landscapes, public spaces, and the built environment	SI 272 – Satisfaction with the beauty of street and buildings in one's neighbourhood	Built environment
		SI 273 – Satisfaction with public parks and gardens	Built environment
		SI 274 – Satisfaction with public spaces	Built environment

Notes:

[a] Cultural objectives listed on the RFSC website are replicated within the RFSC tool, with one exception – "Encourage and value cultural diversity" (website) is converted into two objectives in the tool: "Take specific measures to support language skills for foreigners" and "Promote multicultural exchange between the inhabitants (festivals, fairs, culture exchange, etc.)."

[b] Categories derived from policy documents reviewed in this study.

engagement, and satisfaction, monitored over time, may serve as a proxy), nor do the indicators suggest the roles of culture within broader urban sustainability. The indicators can be reviewed through two lenses: (1) What are the overall characteristics of the suggested indicators? (2) How do the indicators reflect or align with the three cultural dimensions articulated in the policy literature?

Overall characteristics. In the RFSC system, four categories of indicators are evident: on the "supply" side, measures of investments and levels of infrastructure available, focusing on physical assets (listed heritage buildings and land for sports and leisure use); and on the "demand" side, indicators relating to the public's direct use or "engagement" with culture and leisure and their "satisfaction," again focusing primarily on the physical environment of the city, but also considering (with one indicator) their satisfaction with "the level of cultural, recreational, and leisure services" in the community. A general indicator of "civic participation in the local area" complements this but goes well beyond the present scope. No indicator is provided for the objective to promote multicultural exchange among inhabitants. Overall, indicator selection seems influenced by the general data availability.

Alignment with cultural dimensions in urban policy. Viewing the suggested indicators in the light of the three dimensions of culture articulated in "a European vision of the sustainable city" – as a social quality (cultural diversity), as social activities, and as a dimension of the built environment – reveals more starkly the limited extent to which the indicators measure any *cultural* aspects within these dimensions:

- *Social quality / cultural diversity* – This appears to be the least developed area in the RFSC system. Two objectives address a population's cultural diversity, but no indicators are allocated to these objectives, although language services provision could be seen as a yes/no item. However, data relating to cultural, ethnic, or language diversity in a population is commonly available as part of broader city social and demographic data. Although not directly referenced among the selected objectives, this "contextual" information is likely to be integrated in some way into the planning processes using the RFSC system. More explicitly linking these data to the cultural objectives and indicator analyses may be a useful refinement of the system.
- *Social activities* – Culture as a social activity is reflected in the indicators relating to direct employment, use, and "engagement" with the arts. The "use" indicators relate only with municipal library use, which is too narrow to represent this dimension, and the uncertainty in the definition and data available to track "engagement" leaves this dimension vague. Measures of satisfaction with "the level of cultural, recreational, and leisure services" in the community indirectly reflect culture as a social activity, but there is much still to be developed in this area. Overall, it appears that the scope and quality of the indicators in this category may be dependent on the availability of reliable extra-municipal data covering the broad spectrum of activities necessary to capture cultural dimensions of social life.

- *Dimension of the built environment* – Most indicators relate to the physical or built environment, either in terms of provision of "infrastructure" or in satisfaction of the population with different aspects of the city's physical dimensions. The key indicators associated with the two culture-relevant Core Objectives are found in this category, reinforcing its dominance. However, the cultural dimensions of the indicators in this category are absent or largely implicit.

The European Commission and sustainable urban development

Sustainable development is a fundamental principle of the European Union set out in the Treaty, and promoting sustainable urban development is a key element of European Cohesion Policy (EC 2014). The EU "territorial agenda," which ran parallel with urban policy processes, is also linked to urban development. From 2007, with the introduction of the EU *Territorial Agenda* and the *Leipzig Charter*, the European Commission's growing attention to local planning has been noticeable (Campos 2013). By 2011, territorial development was prominent at the EU level, marked by the *Territorial Agenda of the European Union 2020* (EU 2011). Looking forward, greater emphasis will be put on "place-basing of policy," including the coordination and integration of sectorial policies (Campos 2013).

In this context, two recent examples of European Commission initiatives illustrate the nature of efforts to link culture and sustainable urban development. In 2010, a major research initiative, "European Cities of Tomorrow," was undertaken to inform the development of policy for European cities, providing key strategic lines to envision and act on key issues for urban areas. From a more tactical perspective, the *Policy Handbook* was developed in 2012 to raise awareness among local, regional, and national authorities about the potential of cultural and creative sectors in regional and local development, and to help them formulate integrated strategies.

Organized by the European Commission Directorate General for Regional Policy, the "cities of tomorrow reflection process" brought together urban experts and representatives of European cities to think about the future of cities in Europe. Four workshops were organised in 2010, and written contributions were received in the form of issue papers or responses to expert consultations. The exercise contained a SWOT analysis and articulated the challenges and desired trajectories for cities in meeting the objectives of the Europe 2020 strategy – smart, green, and inclusive growth (EC-DGRP 2011).

A European vision of "the cities of tomorrow" sees them as places of

- advanced social progress – e.g. where the elderly can participate in social and cultural life;
- democracy, cultural dialogue, and diversity – with diversity rooted in "culture, identity, history and heritage" incorporating "social diversity and different cultural expressions" (34, 36);

- green, ecological, or environmental regeneration; and
- attraction and engines of economic growth – bringing together "a high quality of life, . . . architecture and . . . functional user-oriented urban space, infrastructure and services, where cultural, economic, technological, social and ecological aspects are integrated in the planning and construction" (11).[2]

A subsequent vision of the *creative city* highlighted the following: "clearly visible" cultural activity produced by a rich and diverse array of established and grassroots groups; the expressive ways in which "inhabitants 'live' their city"; creative use of public space; and an openness to the cultural diversity of newcomers, reflected in new social events and "in a constant political and cultural effort to weave links not only within communities, but also with the rest of the world" (EC-DGRP 2011, 36). From an urban environment perspective, it outlined how culture is "mainstreamed into the provision of public services" such as distinct urban design and way-finding signs and systems; urban infrastructure such as street furniture and street lighting; high visibility of public and community amenities and services in promotion spaces; and visualization of place-specific urban legends and stories by "statuettes, messages or signs on the asphalt" (36). Altogether, it presented an attractive and compelling vision, with some sense of integration, but not one that is readily recognizable as a core component of sustainable urban development.

The *Policy Handbook* was more specifically targeted to integrating culture in regional and local development. The European Commission estimates that between 2007 and 2013 it invested more than six billion euros in cultural infrastructure, cultural heritage, and cultural services through the EU Culture and Structural Funds, and it states that the current challenge is to "further integrate the cultural and creative sectors into regional and local development strategies" (EC-CRD 2014). Towards that end, the *Policy Handbook* (2012) was developed to "better sensitize local, regional and national authorities on the potential of cultural and creative sectors in boosting regional and local development, and help them formulate integrated strategies for these sectors" (EC-CRD 2014).

The *Policy Handbook* notes that the key challenge is how to further integrate cultural and creative sectors into "regional innovation strategies for smart specialization, which . . . will be an ex ante conditionality to access funds" (Working Group 2012, 19). Smart specialization is an innovation policy of the European Commission designed to boost regional innovation through identifying and building on (economic) strengths and "high-value added activities which offer the best chances of strengthening their competitiveness" (EC-RI 2014).

While the *Policy Handbook* includes an array of interesting recommendations for action, few items refer to *sustainable development* or *sustainable cities*. However, *sustainable development* is listed as one of twelve policy areas with which cultural and creative industries (CCIs) have links. A discussion of CCIs and *environmental* sustainability notes that the "specific intervention of CCIs on the environment can contribute decisively to its future preservation" (53). Two Spanish case studies are presented[3] that use cultural approaches to explore and understand the relationship between art and nature/environment, providing spaces for debate and discussion and platforms for exhibitions, publishing, and other activities. While the examples are inspiring, they do not provide much guidance to planners who are looking for strategies to integrate culture within sustainable city/region policies and plans.

Closing reflections

This analysis has shown that cultural dimensions are included in policy/planning frameworks for "sustainable city" building in Europe and have been an ongoing part of policy discussions. However, culture is referenced inconsistently and, in the sphere of urban sustainable development policy, still holds a marginal role. While culture as values is embedded in present European policies, greater attention is needed to the "institutionalized" cultural sector and cultural activities.

This chapter has outlined the main messages and developments from three main groups of actors: European cities, national ministers for urban development in Europe, and the European Commission. From these sources, we observe three key points: (1) culture is recognized politically as an important dimension of local/urban/sustainable development; (2) there is a growing emphasis on holistic and integrated strategies for urban planning and development; and (3) broad planning frameworks and tools to encourage or enable this integration are being built.

Within this policy context, three cultural dimensions are emphasized: (1) the built environment (heritage, architecture, and urban design of public spaces); (2) culture as social activities (in relation to creation, exchange, and access); and (3) culture as social quality (i.e. cultural diversity). An array of suggested objectives and indicators to monitor changes and impacts has been proposed to support planning efforts and investments. However, the selection and coverage of cultural dimensions is uneven, and a more developed conceptual framework is needed to structure the cultural objectives and the indicators. Moreover, in the middle of this picture we find a "black box" at the European level – the issue of *how* to integrate culture into sustainable urban development is not addressed (i.e. operational pathways or methods to support planning practice). Through cultural interventions, policy experimentation, and civic creativity at the local level, avenues of operationalization will be incrementally developed but would be more effectively advanced through facilitated coordination and knowledge networking at the European level.

The issue of integrating culture into sustainable development requires both *conceptual* and *tactical or operational* support and capacity development. To advance this process, two types of city-level tensions must be addressed. The first source of tension relates to the lack of a knowledge base and skills training to support integrated approaches and practices that include culture. The second tension resides within planning systems that continue to support departmental or discipline-specific silos rather than integrated teams. Such "separations" must be viewed from two perspectives: on one side, is "culture" (e.g. cultural officers of a municipality) invited to be part of broader (sustainable development) planning decisions? On the other side, can "culture" (e.g. the municipal cultural department) see itself in these (sustainable development) planning contexts?

From a research perspective, two major gaps require attention. On the conceptual level, the development of an overall framework incorporating the multiple aspects of culture would provide a common starting point from which to consider the various roles of culture in urban development. On the operational level, the issue of *how* to integrate cultural considerations into planning for sustainable urban development should also be investigated. Previous research on topics such as cultural planning and cultural mapping, heritage and historic centre planning, cultural events and urban revitalization, or culture and social inclusion should be revisited and adapted to meaningfully inform a culturally sensitive "urban sustainability" context and practices of urban sustainable development. Complementing this, local experiments in current practice should be examined in a context-sensitive and interdisciplinary manner, as an array of "bottom-up"-informed approaches may illuminate pathways and contribute to the clarification and advancement of these issues.

Notes

1 Running parallel with the urban policy process, the EU "territorial agenda" is also linked to urban development. The *Territorial Agenda of the European Union 2020* (EU 2011) links territorial cohesion with the Europe 2020 strategy and builds on an integrated and cross-sectoral approach. It highlights the importance of improving territorial connectivity and "managing and connecting the ecological, landscape and *cultural* values of regions" (EC-DGRP 2011, 8 [emphasis added]).

2 This dimension also highlights the heritage and architectural value of historic buildings and public spaces for the urban landscape, nurturing places where local residents are given the possibility to "identify themselves with the urban environment" (11).

3 The Beulas Foundation's Centro de Arte y Naturaleza, in Huesca, and the César Manrique Foundation, in Lanzarote, Las Palmas, Spain.

References

Aalborg Charter (1994) Available at: www.sustainablecities.eu/fileadmin/content/JOIN/Aalborg_Charter_english_1_.pdf [Accessed 25 January 2014].

Aalborg Commitments (2004) Available at: www.sustainablecities.eu/aalborg-process/document/ [Accessed 25 January 2014].

Campos, V. (2013) The Cost of Transnational Bonds. Presentation at "Urban (Ir)rationalities: Between Global Dynamics and Local Collective Actions" conference, 23–26 June 2013, Lisbon, Portugal.

Duxbury, N., Jeannotte, M. S., Andrew, C., and Mateus, J. M. (2012) *Culture in Sustainable Communities: Integrating Culture in Community Sustainability Policy and Planning in Canada and Europe*. Report to the Canadian Embassy in Portugal and the International Canadian Studies Association.

European Commission (EC) (2013) The RFSC Conference 2013: Towards a European Model of a Sustainable City, 9 October 2013, Brussels (Belgium). Available at: www.rfsc-community.eu/events/rfsc-conference-2013/ [Accessed 7 February 2014].

European Commission (EC) (2014) Reference Framework for European Sustainable Cities. Available at: www.rfsc-community.eu [Accessed 7 February 2014].
European Commission – Culture and Regional Development (EC-CRD) (2014) Website available at: http://ec.europa.eu/culture/our-policy-development/culture-and-regional-development_en.htm [Accessed 3 February 2014, no longer available].
European Commission – Directorate General for Regional Policy (EC-DGRP) (2011, October) *Cities of Tomorrow: Challenges, Visions, Ways Forward*. Brussels: European Union.
European Commission – Research and Innovation (EC-RI) (2014) Smart Specialization. Available at: http://ec.europa.eu/research/regions/index_en.cfm?pg=smart_specialisation [Accessed 17 February 2014].
European Sustainable Cities Platform (ESCP) (2014) Available at: www.sustainablecities.eu [Accessed 25 January 2014].
European Union (EU) (2007) *Leipzig Charter on Sustainable European Cities*. Available at: www.eukn.org/E_library/Urban_Policy/Leipzig_Charter_on_Sustainable_European_Cities [Accessed 25 January 2014].
European Union (EU) (2008a, November 25) *Final Statement by the Ministers in Charge of Urban Development*. Marseille (France). Available at: http://app.rfsc.eu/userfiles/file/marseille/Final_statement_EN.pdf [Accessed 8 February 2014].
European Union (EU) (2008b, November 25) *Final Appendix to the Final Statement by the Ministers in Charge of Urban Development for Implementing the Leipzig Charter*. Marseille (France). Available at: http://app.rfsc.eu/userfiles/file/marseille/Final_statement_Leipzig_Appendix_EN.pdf [Accessed 8 February 2014].
European Union (EU) (2010) *Toledo Informal Ministerial Meeting on Urban Development Declaration [Toledo Declaration]*. Available at: http://urban-intergroup.eu/wp-content/files_mf/es2010itoledodeclaration.pdf [Accessed 6 March 2014].
European Union (EU) (2011) *Territorial Agenda of the European Union 2020: Towards an Inclusive, Smart and Sustainable Europe of Diverse Regions*. Available at: www.eu-territorial-agenda.eu/Reference%20Documents/Final%20TA2020.pdf [Accessed 6 March 2014].
European Union – Spain Presidency (EU-Spain) (2010) *Toledo Reference Document on Integrated Urban Regeneration and Its Strategic Potential for a Smarter, More Sustainable and Socially Inclusive Urban Development in Europe*. Available at: www.mdrt.ro/userfiles/declaratie_Toledo en.pdf [Accessed 26 January 2014].
UNESCO (2013) *The Hangzhou Declaration: Placing Culture at the Heart of Sustainable Development Policies*. Adopted in Hangzhou, People's Republic of China, on 17 May 2013. Available at: www.unesco.org/new/en/unesco/resources/the-hangzhou-declaration-heralding-the-next-era-of-human-development/ [Accessed 6 March 2014].
Working Group of EU Member States Experts (Open Method of Coordination) on Cultural and Creative Industries (Working Group) (2012) *Policy Handbook on How to Strategically Use the EU Support Programmes, Including Structural Funds, to Foster the Potential of Culture for Local, Regional and National Development and the Spill-over Effects on the Wider Economy?* European Agenda for Culture – Workplan for Culture 2011–2014. European Union. Available at: www.kulturdokumentation.org/eversion/download/policy-handbook.pdf [Accessed 17 February 2014].

6 Culture in development strategies of small and medium-sized European cities

A comparative analysis

Elisabete Caldeira Neto Tomaz

Introduction

In recent decades, a sequence of politico-economic events as well as technological innovations has extensively marked the development of European cities and regions. These events can be explained through an analysis of urbanization processes, which are conditioned by global dynamics and the changes that are occurring under capitalism (e.g. Scott 2007, 2011; Brenner and Theodore 2002, 2005).

The unprecedented level of urbanization that characterizes spatial development should be seen as a result of urban population growth and the physical expansion of urban areas, but also in light of sociocultural changes of everyday life within and beyond the boundaries of the cities (Madsen et al. 2010). More than 75 percent of the European population lives in urban settlements with different features, interdependencies, and development paths (European Environment Agency 2006). Furthermore, increased mobility and connectivity have reformulated the previously restricted understanding of spaces and places and, consequently, development policies, now integrating them in wider global-local, urban-rural relationships.

The decline of classical manufacturing-based industries, coupled with the consequent crisis of the Fordist production system and welfare state regime since the mid-1970s, imposed numerous socioeconomic and environmental challenges to urban development. In addition, technological evolution has supported a new economy characterized by new divisions of labour and modes of production and consumption (Amin 1994; Brenner and Theodore 2002; Jessop 1989; Scott and Storper 1992). In this context, and in a search to convert their urban economies, many governments have approved regulatory arrangements oriented to market rules and commodification. Policy agendas all over the world have been influenced by these ideas, models, and practices, which are mobilized, assembled, and mutated from place to place (McCann 2008, 2011; McCann and Ward 2012; Temenos and McCann 2013). However, the assertion of neoliberal approaches in urban policy has presented diverse intensities and expressions (Brenner and Theodore 2002, 2005; Peck, Theodore, and Brenner 2009; Swyngedouw, Moulaert, and Rodriguez 2002).[1]

In the case of the European Union, despite different conditions and asynchronous institutional changes between member states, there was a common trend towards a market economy and rescaling of the state, upwards to the supranational institutions and downwards to the regional and local levels, which have reconfigured urban governance models (Brenner 2004). The enlargement of local governments' responsibilities and a decrease of financial resources (Nibbering and Swart 2008; Kuhlmann 2006) has led them to seek new forms of management of public resources and affairs with the involvement of private, public, and nonprofit actors.

The ongoing changes mentioned above have also influenced the evolution of European cultural policies, introducing new orientations and concerns in their formulation. One of the recurring subjects, or domain of public intervention, is the moving from a subsidiary vision, focused on support to arts and heritage, to a more managerial attitude and the extension of the policy scope to themes such as urban regeneration or creative industries. Another issue is the centrality of culture in the development policies of territories, in pursuing economic and social gains as well as sustainability goals. Greater awareness of the economic relevance of the symbolic and aesthetic value of goods and services as well as the spaces where they are created, produced, and consumed (Lash and Urry 1994; Scott 1997, 2000) and incremental increases in individual income and leisure time drew the attention of policy makers to the relevance of culture in the revitalization of local economies through the so-called cultural economy (e.g. Scott 1997, 2000; Du Gay and Pryke 2002; Gibson and Kong 2005). The cultural strategies that are formulated and implemented at the local level reflect the rationales associated to development, to evolving thinking about culture, and to the actors' power relations involved in governance processes.

Given the large scale, diversity, and density of relations and resources, the research literature focuses mainly on large cities and metropolises. Most of these publications neglect the culture-based development initiatives experimented on in small cities (e.g. Bell and Jayne 2006; Lorentzen and van Heur 2012) and their capability to change the city's development path. However, these smaller urban centres, particularly those in intermediate and rural regions, have an important role for a balanced European spatial system[2] as mediators between the centre and the periphery, the urban and the rural, and the local and the global. Moreover, the population's increased mobility and connectivity, the potential to promote economic growth in nontraditional sectors, and the demand for healthier lifestyles all emphasize the necessity to understand what is happening in these cities. Consequently, in recent years there is growing interest about how urban centres outside metropolitan areas have included culture as a key component of their development approaches and how these cities position themselves and take advantage of the opportunities and challenges posed by the current socioeconomic and political situation.

Comparing culture development strategies in small and medium-sized cities

The current study developed a comparative analysis of the culture development strategies adopted by four small and medium-sized cities (SMCs) located outside predominantly urban regions (Eurostat 2012).[3] The case studies are located in distinct European countries with different socioeconomic models and related value systems: Finland, Czech Republic, Portugal, and England. A relational approach was adopted, focused on the political processes, the main actors involved (or excluded), and the specificities of the local contexts. I analyzed main strategic documents and conducted more than sixty semi-structured interviews as well as informal conversations with key actors at national, regional, and local levels. Multiple references were examined, particularly images and representations of material reality and urban experiences that were intrinsic to actors' discourses in order to reveal the meanings of culture in urban policies.

Jyväskylä, Central Finland, Finland

In Finland, after 1950, most public services related to the welfare society were assigned or transferred to municipalities, continuing, however, under the tutelage of the state. After 1989, this autonomy was consolidated, the municipalities becoming responsible for the provision of social and healthcare services, education, cultural services, and technical infrastructure. Until the 1960s, cultural policy was essentially an instrument of nation-state formation and the promotion of national identity. Later on, it turned into a sector of the welfare policy, through a wide network of cultural institutions (Pyykkönen et al. 2009). The economic recession in the 1990s questioned this approach and brought out an increasingly neoliberal market trend in public policies. Moreover, at that time, an administration reform was carried out and various regional councils responsible for regional development and planning were created. Culture became, increasingly, a way to achieve economic goals in addition to its intrinsic values. Nationally, the creative industries gained importance in the context of improving international trade, economic growth, and employment opportunities.

Since its expansion in 2009, the municipality of Jyväskylä has encompassed a vast area of 1,171 km^2 of lakes, forests, hills, and small villages in the Central Finland Region. The municipality has a population of 132,062 inhabitants, but the inner city counts only 25,587 residents. As stated by local officials involved in the city's planning and representatives of the Leader-based rural development association,[4] this change is a huge challenge not only in terms of implementing the development strategy but also in reconciling the various interests and identity values of the communities' groups.

The city is renowned worldwide for its iconic architecture by the famous architect Alvar Aalto and for its world-class education, which creates a constant flow of international students. These are complemented with a lively cultural scene involving theatre companies, orchestras, and a set of cultural institutions (e.g. the Jyväskylä Art Museum, the Craft Museum of Finland, the Alvar Aalto Museum, among others). The most significant cultural branches in Jyväskylä are printmaking and photography, which benefit from local infrastructures such as the Centre for Printmaking. Cultural activities are supported mainly through public funds from the municipality and the national and regional arts councils but also the Finnish Cultural Foundation and the Nordic Fund.

One of the images that best describes the city is its compact urban centre, surrounded by an extensive natural environment that supports local aspirations of connecting quality of life and well-being to innovation technology. The creative economy and cultural entrepreneurship are key topics promoted in national programmes and by the Central Finland Region, which recognized the need to raise awareness of these issues at the local level.

Like other sectors, culture is understood in a transdisciplinary way, connecting the areas of arts, wellness, and entrepreneurship. Local and regional authorities stressed culture's special role to addressing the city's social problems. The elderly and the young are main targets of the city's activity, besides artists' support (emphasized by the Cultural Services Unit). The Jyväskylä Artists' Association member that was interviewed (an independent artist), although agreeing with having a role in community development, argued that they should have a better social protection and greater participation in decision making. New strategies have been tried by the municipality – namely, in the Kangas renewal area – to improve collaboration and participation in development through social media and face-to-face meetings. Another example is the Veturitallit Centre that promotes cultural activities for youngsters and encourages them to express their opinions about urban problems. Moreover, the strength of voluntary associations and citizens in community development is perceptible in many fields.

The region has benefited from EU programmes in terms of structural funds but also in knowledge exchanged through diverse networks. These instruments had encouraged the relation between urban and regional development, on one hand, with social cohesion and cultural policies, on the other (Mitchell and Heiskanen 2011).

Český Krumlov, South Bohemia, Czech Republic

Contemporary Czech urban policy is the result of the profound transformations that took place in many post-socialist countries that joined the EU. The transition to a pluralist democracy and market-based economy gave more autonomy to local governments, adding new responsibilities. In the late 1990s, local politicians became more aware of the importance of urban planning and the necessity to deal with sustainability issues. Since 2009, the economic crisis and state budget deficits have contracted support for cultural projects.

Český Krumlov is situated in the South Bohemian Region, classified as a predominantly rural area. This medieval town grew due to its privileged location on an east-west communication route, the Vltava River, flourishing as an important craft and trade centre and as a regional cultural reference. The remarkable castle, well-preserved Renaissance and Baroque burgher houses, and a scenic river, complemented by the surrounding attractive natural landscape of the Šumava foothills, leave a strong impression and sense of uniqueness.

The inclusion of the city centre on the UNESCO World Heritage List in 1992 was an opportunity to initiate urban renewal processes. A strategic development plan was initially developed with the help of external consultants and later elaborated further by representatives of municipal authorities, major institutions, and companies. As part of UNESCO requirements, the historic centre has a specific action plan that states the significance of the heritage for city development. To manage the properties in the historic centre, the Český Krumlov administration established a company, Český Krumlov Development Fund, and a tourism department, Destination Management, to ensure the implementation of the City Council's strategy. The castle administration also plays a significant role in the local development strategy and cultural activities.

As a result, the city became more attractive for investors, property developers, and visitors. Currently, it is one of the most popular destinations in the Czech Republic. To revive the historical centre, a range of cultural events were developed, such as the "Five-Petalled Rose Celebrations" and the "Festival of Baroque Arts" involving many local actors (e.g. Foundation of the Baroque Theater, Chamber Music Festival Foundation, and prestigious cultural amenities such as the Egon Schiele Art Centre). The goal is to increase the number of tourists beyond the main season, as a source of revenue for socioeconomic development projects. For that reason, a range of leisure events and facilities and a strong marketing strategy are promoted.

Most of the cultural projects, even those that are privately initiated, are developed with the financial support of the city of Český Krumlov, the central state, and the South Bohemian Region. Cultural activities are strongly dependent on public funds or other sources such as European programmes, the UNESCO Heritage Programme, and EEA and Norwegian Grants. Cultural projects benefit from the existence of active cross-border collaborative ties.

However, the city development approach has generated some conflicts and divergent positions. A certain level of mistrust between the local political elite and some of the most influential cultural actors is perceptible. Another concern of local cultural actors is the difficulty in establishing partnerships and the mobilization and participation of civil society. The main problems mentioned by interviewees emerge from the large flow of tourist groups and associated problems such as noise, the number of stores with standard- and lower-quality products, a decrease in residents living in the downtown, and the rising cost of goods and services. The major challenge for the current cultural policy of the city is to maintain *the genius loci* that once attracted artists and international associations, with initiatives preserving the collective memory (like in the Elementary Art School or Fotoateliér Seidel Museum) and improving the quality of life of disadvantaged groups.

Óbidos, Central Region, Portugal

In Portugal, the National Spatial Development Policy Programme (2007–2025) sets the framework for development priorities at the national and local level. In recent years, the country witnessed a decentralization policy along with a decrease in state budget transfers and a reduction of local tax revenues. Consequently, local governments have adopted different organizational structures, like public and inter-municipal companies, to provide public, including cultural, services. After the period of the dictatorship, national cultural policy focused largely on infrastructure creation and renovation, organized in networks through partnerships between central and local government (Silva 2004). Local authorities are also responsible for promoting access to culture and art education, supporting identity-related initiatives, and supporting internationalization through participation in networks of production and consumption (Silva 2007).

The small municipality of Óbidos, with only 11,772 inhabitants, is located a close distance from the capital. It covers an area of 142 km^2 and is classified as a predominantly rural region. Óbidos is distinguished by its iconic image: the medieval walled village with narrow streets and white traditional houses. It is also characterized by a unique natural landscape featuring a lagoon and an extensive coastline. Local identity is shaped by its historic and noble past alongside rural traditions and lifestyles, both reflected in local heritage. To conserve the historic buildings in 1911, the castle was classified as a National Monument, and this was extended to the entire walled town centre in 1951.

In 2001, while seeking to reverse the process of social and economic decline, the elected mayor initiated a development strategy to convert the rural-based economy through culture and creativity, supported by a strong marketing strategy: "Creative Óbidos." Two municipal companies were created to manage local initiatives, and diverse public-private partnerships were promoted with higher education institutions, training companies, and business associations.

From the beginning, the local strategy was supported by a lively agenda of thematic events such as "International Piano Week" or the popular "International Chocolate Festival," which contributed to boosting the city's image and tourism. This was underpinned by a set of cultural facilities (e.g. Network of Museums and Galleries), environmental sustainability programmes, and a new educational system that uses creativity as a way of building new pathways and innovative approaches to community development (emphasized by municipality representatives and recognized abroad).

A strategic priority to attract creative and skilled individuals was realized through a set of financial incentives and investments to support creative business (i.e. incubator spaces). Recently, an urban regeneration programme was launched that will provide studios, co-working spaces, and live-work houses for national and international artists, designers, researchers, and so forth. In order to reinvent themselves, local decision makers foster a creative atmosphere that seeks to boost the territory beyond its formal limits. Internationalization through participation in exchange networks, usually through European programmes, has served as an opportunity to enhance the city's visibility and recognition and, more importantly, to acquire critical mass and knowledge concerning their development specialization in creativity.

Furthermore, the proximity between local actors, direct contact with local decision makers, and the mobilization of foreign agents have all contributed to trying new possibilities. A good example is the Óbidos Literary Town, a project that involves the installation of a bookshop network in this small settlement, in a diversity of spaces (churches, museums, etc.) along with meetings, workshops, and festivals around literature. The project counts on the participation of private entrepreneurs and the local municipality and is seen as a way to consolidate its sustainability.

However, a major challenge is to combine a contemporary and cosmopolitan imaginary in a rural territory. The discourse of cultural and political elites is unintelligible to a substantial part of the community, which creates resistance and obstacles to legitimate action and wide-ranging development. This is partly covered by sociocultural projects like Odesign, which links young designers and seniors to combine design and creativity with traditional techniques and skills.

As the museum director declared, it is necessary to reinforce the genuine cultural component, the relationship between tangible and intangible heritage, the built and natural environment, which gives it a distinctive character. Indeed, one of the biggest criticisms made by local residents is the predominantly tourism-oriented nature of services, which have resulted in numerous visitors that disturb their daily life, the commodification of local values and practices, and the conversion of the community into a "scene."

York, North Yorkshire, United Kingdom

Over the past thirty years, in the UK, institutional reforms were carried out that impacted spatial development and the formulation of cultural policies. In particular, the reorganization of local government led to reallocating certain public services to private sector and nonprofit organizations through outsourcing, public-private partnerships, and other policy instruments. More recently, regional development institutions were abolished, to be replaced by Local Enterprise Partnerships. The British cultural policy – founded on the "arm's length principle" that is characterized by noninterference, limited support from the state, and the rising intervention of civil society and private actors (Gattinger and Saint-Pierre 2008; Menger 2010) – has shifted to a market-oriented and managerial approach where culture acquires an increasingly instrumental role. In this framework, UK creative industries policy developed a huge influence that extended around the world. Local authorities had an essential role in supporting arts and crafts and in preserving local heritage not only via their culture budget but also indirectly through regeneration agencies and other statutory and strategic agencies. Other funding sources are primarily DCMS grants, Arts Council of England and National Lottery programmes, private sponsors, and EU programmes.

The City of York is a unitary authority of 198,000 inhabitants, a nonmetropolitan district located less than two hours from London, in the intermediate region of North Yorkshire County. Within an area of about 272 km^2 there are distinct localities, suburbs, and villages surrounded by a large rural landscape. Recently, York has moved from a prosperous economy based on chocolate manufacturing and railway-related industries to one dependent on the tertiary sector, including mostly tourism, health, education, and other public services.

With a unique atmosphere characterized by two thousand years of urban settlement, the city centre has well-preserved archaeological sites, ancient monuments, and notable archives that reflect the history and culture of England and Europe. "This is its character," as various actors emphasized. Given this legacy, the Cabinet Leader intends to place a new accent on innovation grounded in the city's history as a starting point for discussing what the city wants to be.

Culture only recently began to be openly referred to in strategic documents. In 2004, the Local Strategic Partnership – incorporating representatives of the public, business, and voluntary sectors – developed the first cultural strategy after an extensive public consultation. Subsequently, a cultural partnership was formed, York@Large, and the City Council pursued one of the main dimensions of the city's strategy, named "York, a city of culture," distinguishing it for its inclusive, lively, and active profile. In 2011, given its awareness of the importance of place marketing, the City Council and a group of stakeholders submitted a bid to integrate the city onto the UNESCO World Heritage List. After the first attempt failed, York launched a second bid to become a UNESCO City of Media Arts, in an effort to transform creative industries into a driver for urban development and to link culture, innovation, and technology.

The city also has a set of cultural institutions, ranging from museums and interpretation centres to cultural charities. Many of them combine cultural and educational experiences alongside commercial activities. Moreover, an extensive agenda of festivals was created, supported by local organizations, recreating an imaginary based on local identity and history but also approaching contemporaneity. The visitor economy is central in the local development strategy, representing seven million annual visitors, 20,200 jobs, and a thriving industry of £606 million in 2012 (Visit York 2014).

Challenges to urban planning are balancing preservation and economic growth concerns in a dense and compact centre, the city's proximity to large labour markets, a lack of office spaces in the centre, and reduced financial capacity. In turn, some cultural institutions highlight an excessive focus on tourism and white-collar employment and the high cost of dwellings in the inner city, which pushes the lower-income classes to outside wards. Their key challenge is to combine economic objectives with an emphasis on community development.

Key findings

The SMCs in this study accept culture as a key component of their development strategies. Although using different approaches and understandings, local authorities recognize the potential of culture as a way to address urban problems such as the decline of traditional industries and, consequently, the growing unemployment and lack of financial resources.

The four cities attempt to explore new ways of development, seeking to specialize or diversify the local economy based on its distinctive character and history. Thus, Jyväskylä and York's development are rooted in their industrial past but also in the solid reputation built in the educational field in connection with innovation policy, allowing them to compete in attracting businesses and international students. In Český Krumlov and Óbidos, built and natural heritage anchors the development of events and tourism, while the towns are also trying to create new possibilities to diversify the local economy. For Český Krumlov, the UNESCO classification and European funds marked the turning point, while Óbidos linked culture and education to the creative economy and turned it into an engine of development. The creation of a vibrant and attractive image for mobile knowledge workers is a strategic issue in York, as well as in Jyväskylä and Óbidos, but in the latter two this is aligned with the enhancement of quality of life and well-being.

The four local authorities implemented similar types of interventions: from the creation of a rich cultural agenda and cultural venues to the regeneration of old quarters and abandoned buildings in new artistic and creative spaces. But they mobilized different interests and particular relations of meaning. Despite the political rhetoric, they advocated an economic rationale in culture development strategies beyond social concerns. However, a community development focus depends largely on the national framework and actors' power relations (see Table 6.1). For example, in Jyväskylä, local discourses retain long-standing principles of Finnish cultural policy: to ensure cultural participation as a way to promote health and psychological well-being. The English case can be seen in the context of the British neoliberal policy agenda, where culture is regarded as a driver of growth and its institutions as a powerful instrument for renewal projects and partnerships. Furthermore, cultural actors highlight community development as an inalienable aim, making them sustainable through commercial activities. In small communities like Český Krumlov and Óbidos, local associations and education institutions are crucial actors for community engagement in development strategies. In the post-socialist case, there are still perceptible discursive practices of a strong and bureaucratic government and a traditional cultural attitude of valuing high culture and expressions of collective identity. The bureaucratic model of national policy is also an impediment for the municipality of Óbidos, which continues to pursue greater autonomy from the state.

Public authorities have introduced new governance models, creating agencies and departments to manage cultural urban policies and to establish public-private partnerships and inter-sectoral collaborations. A singular situation is the role of York Trusts in cultural and urban planning (e.g. York Civic Trust for conservation regulation), which are nonprofit, successful models in the cultural field. Jyväskylä is the only municipality that has a cultural unit and a regional body with influential urban cultural policies. Nevertheless, local institutional capacity – especially the strong commitment and proactive role of local offices plus the resilience and capacity of nongovernmental actors – was decisive to test new development approaches. The proximity between stakeholders and direct contact with decision makers is an advantage that is often used as a way of solving multiple coordination and legitimation-related problems. Flexibility, the small size of the projects, and the incorporation of motivational elements are critical to its sustainability, especially in small towns like Óbidos. Another key aspect is

Table 6.1 Cultural development approaches

Development agenda and culture				
Approach	*Rationales*	*Objectives*	*Key interventions*	*Leading actors*
Economic development	• To create wealth and jobs • To increase resilience and competitiveness • To diversify local economy	• Support entrepreneurship and cultural production • Improve visitor economy • Stimulate inward investment • Promote cultural consumption • Enhance a positive place image • Retain graduates and skill workers • Promote spill-over effects	• Infrastructures for cultural/creative business • Fiscal incentives and start-up support • Place branding and tourism services • Cultural/creative clusters • Events and festivals • Arts venues • Heritage sites • Fairs and congresses • Renewal projects • Grants for research projects • Public services privatization • Retail and leisure areas	• Public authorities • Development agencies • Business • Political experts
Community development	• To create more equitable, healthy, and diverse communities • To promote high quality of life and well-being	• Provide access to culture and art education • Enrich civic participation and community engagement • Reduce social risks • Improve social cohesion • Stimulate collective identity • Promote cultural diversity and inclusion	• Cultural institutions funding • Artists grants • Education and art programmes support • Cultural/social centres • Low-rent cultural facilities • Multicultural and community events • Forums and public consultation • Voluntary programmes • Neighbourhood renewal	• Cultural institutions • Civic associations • Artists • Educational institutions

the involvement of the community in initiatives beyond attracting visitors and businesses, such as the Óbidos Creativity Factory programme that brings together recreational, artistic, and cultural initiatives aimed at promoting creativity in schools and community.

The growth of tourism in historic and compact centres such as in York, Český Krumlov, and Óbidos represents a great challenge for the design of urban policy, especially in order to avoid gentrification processes. In contrast, in Jyväskylä, the enlargement of the municipality with the inclusion of vast rural areas has brought great difficulties for planning to include such diverse interests.

Local stakeholders think the main challenge to the sustainability of cultural strategies in coming years is to deal with reduced funding and private sponsorship. This depends largely on evaluating the direct impact of investment and simultaneously discovering new sources of funding in order to retain some initiatives that meet the community's expectations and address its problems. The contribution of culture to economic and tourism growth in these cities is typically valued more than its role in community development.

Final remarks

The macro processes identified in the first part of this chapter are important to understand local politics and, in particular, how cultural policies follow the overall development of wider social and economic trends and how local discourses and practices are influenced by ideas and models circulating in transnational networks where urban elites are involved. Furthermore, despite the growing autonomy of local authorities, cities' development agendas are also preconditioned by national strategies and EU policy and structural funds, although in different intensities.

Culture as a field of public intervention has become a major driver of urban development and an instrument of legitimation for political actions of these SMCs. In general, their strategies denote an overlapping of the economic agenda and the formulation of cultural policies while simultaneously seeking to achieve social goals through community-oriented initiatives and educational programmes. Apart from strong political commitment and an entrepreneurial attitude, local authorities established new modes of governance, including informal arrangements, in order to implement the necessary changes. At the same time, the political processes reproduce the political-administrative settings and trajectories of development of each city. Moreover, discursive strategies are built through multiple references and meanings, and influenced by ideas, experiences, and material representations of places that are mobilized in governance processes (Healey 2007). Discourses and practices reproduce actors' interests and power relations that are constantly negotiated, and where some groups and themes are excluded.

Despite the general predisposition to evaluate the success of culture-led urban policies in accord with a city's size and position in urban hierarchy, local actors underlined large cities' disadvantages such as traffic congestion, social exclusion, environmental problems, and so forth, in contrast with community proximity, greater flexibility, a healthier quality of life, and easy mobilization, among other features of smaller cities. Increased connectedness and mobility have augmented these cities' functions and possibilities, especially those in intermediate regions, which transforms these cities into laboratories capable of more quickly implementing changes.

In general, culture is understood as a tool to generate an attractive and distinctive environment for people and organizations, and as a resource for the development of production and consumption activities in the context of the cultural economy. But although culture is considered an economic asset and its economic impacts, especially spillover effects, often serve to justify public funding of the sector, cultural and artistic institutions still tend to feel marginalized from decision-making processes and increasingly vulnerable to external pressures.

The sustainability of these initiatives depends largely on their ability to develop business and fundraising expertise and to adjust their contents and programmes to national and European initiatives and funders. In small towns where the cultural budget is reduced, the priority is given to large organizations and popular events. However, part of the cultural sector also learned to benefit from cultural policies linked to development projects.

A strong sense of community and identity, capable of legitimizing cultural policy actions and fostering an innovative environment supported by networks and mobilizing agents, seems crucial to lever the necessary changes that these cities pursue. Regardless of the size of the city, long-term sustainability depends on achieving a balance between community values and economic purposes.

This paper sought to explore some assumptions regarding the development of SMCs. It is essential to understand that there is no "one-size-fits-all" model. Further, it is necessary to consider not only the structural factors but also political agency and the dynamics that affect local political processes and, subsequently, make feasible the development strategies of small towns through their distinctive qualities of place and their talents – in short, development *of their culture*.

Notes

1 For a typology of neoliberalization phases in urban policy, see Peck and Tickell (2002).
2 According to the new urban-rural typology for NUTS 3 regions, only 9.9 percent of the EU territory is designated as predominantly urban regions, representing 42.4 percent of the European population (Eurostat Regional Statistics Team 2013).
3 For more detail on this typology, see Eurostat Regional Statistics Team (2013).
4 Leader local action groups (LAGs) are associations that develop rural areas by funding local rural development projects and supporting local enterprises. There are fifty-six LAGs active in Finland (Finnish Leader 2012).

References

Amin, A. (1994) Post-Fordism: Models, fantasies and phantoms of transition. In A. Amin (ed.), *Post-Fordism: A Reader* (pp. 1–39). Oxford: Blackwell.
Bell, D., and Jayne, M. (2006) *Small Cities: Urban Experience Beyond the Metropolis*. London: Routledge.

Brenner, N. (2004) Urban governance and the production of new state space in western Europe. *Review of International Political Economy*, 11(3), 447–488.

Brenner, N., and Theodore, N. (2002) Cities and the geographies of "Actually Existing Neoliberalism." *Antipode*, 34(3), 349–379.

Brenner, N., and Theodore, N. (2005) Neoliberalism and the urban condition. *City*, 9(1), 101–107.

Du Gay, P., and Pryke, M. (2002) Cultural economy: An introduction. In P. Du Gay and M. Pryke (eds.), *Cultural Economy: Cultural Analysis and Commercial Life* (pp. 1–19). London: Sage.

European Environment Agency (2006) *Urban Sprawl in Europe: The Ignored Challenge*. Report No 10. Luxembourg: Office for Official Publications of the European Communities.

Eurostat (2012) *Eurostat Regional Yearbook*. Available at: http://epp.eurostat.ec.europa.eu/cache/ITY_OFFPUB/14_FOCUS_ON_TERRITORIAL_TYPOLOGIES_RYB2012/EN/14_FOCUS_ON_TERRITORIAL_TYPOLOGIES_RYB2012-EN.XLS [Accessed 4 May 2013].

Eurostat Regional Statistics Team (2013). Urban-rural typology update. Available at: http://epp.eurostat.ec.europa.eu/statistics_explained/index.php/Urban-rural_typology_update [Accessed 8 December 2014].

Finnish Leader (2012) Finnish Leader scheme a success story. Available at: www.rural.fi/en/index/leader.html [Accessed 8 December 2014].

Gattinger, M., and Saint-Pierre, D. (2008) Can national cultural policy approaches be used for sub-national comparisons? An analysis of the Québec and Ontario experiences in Canada. *International Journal of Cultural Policy*, 14(3), 335–354.

Gibson, C., and Kong, L. (2005) Cultural economy: A critical review. *Progress in Human Geography*, 29(5), 541–561.

Healey, P. (2007) *Urban Complexity and Spatial Strategies*. London: Routledge.

Jessop, B. (1989) Neo-conservative regimes and the transition to post-Fordism. In M. Gottdiene and N. Komninos (eds.), *Capitalist Development and Crisis Theory: Accumulation, Regulation, and Spatial Restructuring* (pp. 261–299). Basingstoke: Macmillan.

Kuhlmann, S. (2006). Local authorities between the state and the market: An international comparison of local government systems and reforms. *German Journal of Urban Studies*, 45(2), 84–102.

Lash, S., and Urry, J. (1994) *Economies of Signs and Space*. London: Sage.

Lorentzen, A., and van Heur, B. (2012) Introduction: Cultural political economy of small cities. In A. Lorentzen and B. van Heur (eds.), *Cultural Political Economy of Small Cities* (pp. 1–14). Abingdon and New York: Routledge.

Madsen, M. F., Kristensen, S. B. P., Fertner, C., Busck, A. G., and Jørgensen, G. (2010) Urbanisation of rural areas: A case study from Jutland, Denmark. *Geografisk Tidsskrift-Danish Journal of Geography*, 110(1), 47–63.

McCann, E. (2008) Expertise, truth, and urban policy mobilities: Global circuits of knowledge in the development of Vancouver, Canada. *Environment and Planning* A, 40(4), 885–904.

McCann, E. (2011) Urban policy mobilities and global circuits of knowledge: Toward a research agenda. *Annals of the Association of American Geographers*, 101(1), 107–130.

McCann, E., and Ward, K. (2012) Assembling urbanism: Following policies and 'studying through' the sites and situations of policy making. *Environment and Planning* A, 44(1), 42–51.

Menger, P. (2010) *Cultural Policies in Europe: From a State to a City-Centered Perspective on Cultural Generativity*. Tokyo: National Graduate Institute for Policy Studies. Available at: www3.grips.ac.jp/~pinc/data/10-28.pdf [Accessed 7 June 2014].

Mitchell, R., and Heiskanen, I. (2011) *Country Profile: Finland. Compendium of Cultural Policies and Trends in Europe*, 13th ed. Available at: https://www.jyu.fi/ytk/laitokset/yfi/en/studies/perspectives-on-finnish-society/external-links-books/books/COMPENDIUMfinland_062011.pdf [Accessed 14 February 2015].

Nibbering, J. W., and Swart, R. (2008) *Giving Local Government a More Central Place in Development: An Examination of Donor Support for Decentralisation*. Available at: www.delog.org/cms/upload/pdf/giving_local_government_a_more_central_place_in_development.pdf [Accessed 8 December 2014].

Peck, J., Theodore, N., and Brenner, N. (2009) Postneoliberalism and its malcontents. *Antipode*, 6, 1236–1258.

Peck, J., and Tickell, A. (2002) Neoliberalizing space. *Antipode*, 34, 380–403.

Pyykkönen, M., Simanainen, N., Sokka, S., and Simanainen, N. (2009) *What about Cultural Policy? Interdisciplinary Perspectives on Culture and Politics*. Jyväskylä: Jyväskylän yliopisto.

Scott, A. (1997) The cultural economy of cities. *International Journal of Urban and Regional Research*, 21(2), 323–339.

Scott, A. (2000) *The Cultural Economy of Cities: Essays on the Geography of Image-Producing Industries*. London: Sage.

Scott, A. (2007) Capitalism and urbanization in a new key? The cognitive-cultural dimension. *Social Forces*, 85(4), 1465–1482.

Scott, A. (2011) Emerging cities of the third wave. *City*, 15, 289–381.

Scott, A. J., and Storper, M. (1992) Industrialization and regional development. In A. J. Scott and M. Storper (eds.), *Pathways to Industrialization and Regional Development* (pp. 3–15). New York: Routledge.

Silva, A. S. (2004) As redes culturais: Balanço e perspectivas da experiência Portuguesa, 1987–2003. In M. L. L. Santos (ed.), *Públicos da Cultura* (pp. 241–283). Lisboa: Observatório das Actividades Culturais.

Silva, A. S. (2007) Como abordar as políticas culturais autárquicas? Uma hipótese de roteiro. *Sociologia – Problemas e Práticas*, no. 54, 11–33. Lisboa: Centro de Investigação e Estudos de Sociologia (CIES).

Swyngedouw, E., Moulaert, F., and Rodriguez, A. (2002) Neoliberal urbanization in Europe: Large-scale urban development projects and the new urban policy. *Antipode*, 34(3), 542–577.

Temenos, C., and McCann, E. (2013) Geographies of policy mobilities. *Geography Compass*, 7(5), 344–357.

Visit York (2014) *Facts and Figures on Tourism in York 2012 Update*. Available at: http://mediafiles.thedms.co.uk/Publication/YK/cms/pdf/07-research-Keyfacts2012.pdf [Accessed 8 December 2014].

7 Cultural policy making by networking

Local cooperation and global competition in small and medium-sized Italian cities

Davide Ponzini

Introduction

Can a network of small and medium-sized cities integrate their cultural service provision or promote innovative cultural policy to the extent of being recognized at a national or even international level? How can such a network lever the legacy of past urban regeneration in a period of economic stagnation? Starting from these questions, this chapter highlights international attempts and examples of this trend in Europe and comments on current contributions in academic and cultural policy literature. More specifically, the chapter addresses the original attempts and tentative solutions in an Italian case study: since the mid-2000s, seven medium-sized municipalities in the Northern Milanese subregion have started to come to terms with common problems of potential redundancy in local cultural services. They have also faced the problems of promoting cultural initiatives in a sustainable way and representing them in a metropolitan region that is dominated by and centred in the capital city of Milan. These problems were addressed by trying to assemble, rescale, and reprogramme part of the governance of cultural service provision, in artistic and creative production and other policy fields. At the same time, the network has promoted outstanding projects in order to compete globally, but on the way it has encountered significant political and financial difficulties and failures.

These issues are common to many parts of Europe. This chapter suggests that a number of sustainability and managerial problems emerging in cultural policy networking can be better understood if connected to urban development processes and the related economic interests.

Planning and policy making issues of small and medium-sized cities in Europe

Historically, the primary interest and research conducted in the field of urban and regional governance and policy making has focused on large cities or metropolitan regions. Today, we know that the variety of spatial and sociopolitical conformations is definitively higher, and we rarely have an appropriate lexicon and set of analytics to face the rising problems (Soja 2000). These territories often combine megacities, metropolitan centres, and a set of smaller territorial entities. In this broad picture, interest in small and medium-sized cities (SMCs) has started to rise in the European debate due to the high environmental, economic, and cultural relevance of such types of settlements. As well, some sociopolitical features in European cities show peculiarities that derive from a long history of small-scale urban organization (Bagnasco and Le Galès 2000; Le Galès 2002; Le Galès and Therborn 2010).

According to one of the groundbreaking research studies on small and medium-sized European towns (ESPON 2006), one can find three major ways of approaching the topic: *morphological* (according to the shape and size of human settlements); *functional* (according to the social and economic relationships existing among the settlements of one specific area); and *administrative* (according to given quantitative criteria such as population thresholds, or the institutional powers of the relevant local authority). Such diverse approaches to the definition of SMCs coexist in most European countries.

In the European continent, the topic of SMCs seems relevant because about half of the urban population lives in towns and cities of one hundred thousand or less inhabitants (European Commission 2011). The debate regarding the gravitation of these centres (however they are defined) around metropolitan cores is vast and articulated (see, for example, the ESPON TOWN research project, Servillo et al. 2012). At the same time, the institutional and political functioning of smaller realities (e.g. social capital and place-based tacit knowledge) can vary significantly and requires contextual attention (Kearns and Paddison 2000; Laurila 2004; Bell and Jayne 2006). The limited concentration of economic and fiscal means in space due to the size of cities sometimes affects the provision of public services and, in particular, of specialized cultural services. The latter – given the current retrenchment of the welfare state – often require the creation of larger pools of cities and actors in order to generate innovative organizational forms and social networks for sustaining and eventually expanding cultural services and policy. According to the existing literature, this situation is counterbalanced by the ability and often the tradition of SMCs to network among themselves, to integrate public policy, and to foster economic and social innovation and cooperation (often facilitated by the existence of close-knit networks and a limited number of participants) in order to compete at a higher scale (Moulaert, Martinelli, Swyngedouw, and Gonzalez 2005; Baker 2006; Moulaert, MacCallum, Mehmood, and Hamdouch 2013). On many occasions in the cultural policy field, this has meant competing with other (nearby) metropolitan realities rather than with other SMCs in the same region. In general terms, we have today only a limited understanding of these conformations both in social and in spatial terms, despite the fact that they can have relevant impacts in cultural governance and economic spillover effects (INTELI 2011; Lorentzen and van Heur 2013).

One can easily see this trend in Europe by considering recent examples. In the last decade, more and more groups of cities have partnered in the provision of public services or even have taken the lead in the candidacy and (if awarded) in the management of a European Capital of Culture (ECoC) initiative – for example, "Essen for the Ruhr" in 2010, Lille Metropole in 2004, or the Marseille-Provence 2013 experience (Ponzini and Ruoppila 2011). Antonio Talone (2011) reports that, in Pecs 2010, the regional level of programming had important results for developing ECoC activities and the relationships among cultural operators and stakeholders in the region thanks to the autonomy from the central government in the programme management. Paris and Baert (2011) offered an interesting image of Lille's development trajectory (since the 1980s) in which the 2004 ECoC programme was nested. Lille was interpreted in its wider metropolitan area with reference to European infrastructures such as the Channel Tunnel and its spatial development strategy, as in the case of Euralille. Claire Bullen (2011) discussed the programme for the Marseille 2013 ECoC. The change of scale from the city to the urban region implied that the cultural initiatives were programmed in the framework of a wider vision for urban development. In this case, the urban development images were intended to impact not only the perception of public officials, political leaders, and social and cultural operators in the region but also local communities.

Issues in Italian urban regeneration policy and cultural programming

The Italian context seems inclined to these experiments of networking. The Veneto Region's failed candidacy for the European Capital of Culture 2019 is just one among a broad set of examples that can be highlighted. The number of policies for promoting so-called cultural districts can be seen as an appropriate example (Ponzini, Gugu, and Oppio 2014). The Lombardy Region has been experimenting with local systems for cultural preservation. In the early 2000s, the Accordo Quadro di Sviluppo Territoriale (AQST) Magistri Comacini in the Lake Como area attempted significant innovation and the explicit creation of a cultural district among a dozen SMCs. In a similar vein, in 2005 the Veneto Region started a systematic promotion of spatial systems for cultural preservation and production throughout the regional territory: Distretti Culturali Evoluti (DiCE) and one of the largest bank foundations promoted another strand of financing in Lombardy (Ponzini, Gugu, and Oppio 2014).

Several scholars and policy makers assumed that cultural heritage and the arts were "engines for economic development." Due to the increasing importance of immaterial and cultural production and the existing cultural heritage and supply in the national context, this appeared to be a natural route to take, fostering the creation of cultural facilities and services. In Italy one must recognize a long-term coupling of cultural policy making with urban regeneration and real estate appreciation, especially including public-private programming and tools (Ponzini 2008, 2010b). Since the end of the 1990s, these new tools and the increase in public-private partnerships in the cultural sector made several authors and policy makers believe that the regional level was the appropriate place for such practices (Petraroia 2003), where the joint negotiation of urban development projects, and the creation and management of cultural facilities and their complementary economic activities (e.g. cafeterias and bookshops) and social services, have been possible.

In facing such innovations and uncertainty, urban regeneration projects have sometimes proceeded through incremental pathways. A central question has been the tendential economic inefficiency of cultural services, typically to be managed and sustained prevalently with local public resources. Today it is possible to highlight a number of critical points, especially in connecting real estate appreciation and the management of cultural facilities. On one hand, the creation of cultural facilities is justified by the alleged positive externalities and their importance in urban regeneration and economic development. On the other hand, in the face of the unresolved ambiguities of public-private programming, one can question whether the cultural facilities and the financing required for their management can be interpreted as a by-product of real estate appreciation and urban regeneration, induced by the sharing of public programming discretion with private parties (Codecasa and Ponzini 2011).

At the same time, one must recognize a dramatic retrenching of public administration in the field, especially when it comes to initiatives of networking or individual investments in small urban areas. To provide one concrete example and the background for the case study of the North Milanese network initiatives, we can consider the programmes of the Province of Milan.

It is extremely important to understand the role played by the Province of Milan and highlight the territorial visions through which it promoted significant cultural networking among SMCs in recent years. The programme "Metropòli" is probably the most interesting since it allows one to scan the implications of the weakening of the province's initiatives in the metropolitan area. Metropòli was started in 1995 with the aim of enhancing the historical and artistic heritage in the outskirts and hinterland of the metropolitan core of Milan. Initially, the programme was intended to promote activities and events to bring new audiences to places of interest and cultural landscapes outside of the capital city. Over time, Metropòli has become a way of aggregating more or less stable groups of neighbouring small and medium-sized municipalities. The municipalities networked into geographically adjacent groups (labeled as "Poli" – i.e. "Poles"; see Figure 7.1) in order to programme cultural initiatives of different types. This generated the conditions for networking and promoting a number of other joint initiatives by more or less the same members of one "Pole."

In 2004, the attempt to make each of these Poles self-sustaining aimed at the creation of some sort of cultural districts led by the municipalities. The attempts and bottom-up initiatives that were generated could not match the small annual funding of the programme that the Province granted. Some Poles have been able to attract local resources and survived through time (e.g. Groane Pole); others have focused on individual events without consolidating the joint cultural programme. The Metropòli programme ceased to provide resources in 2013, and in most cases the programmes have been discontinued. One must notice that the

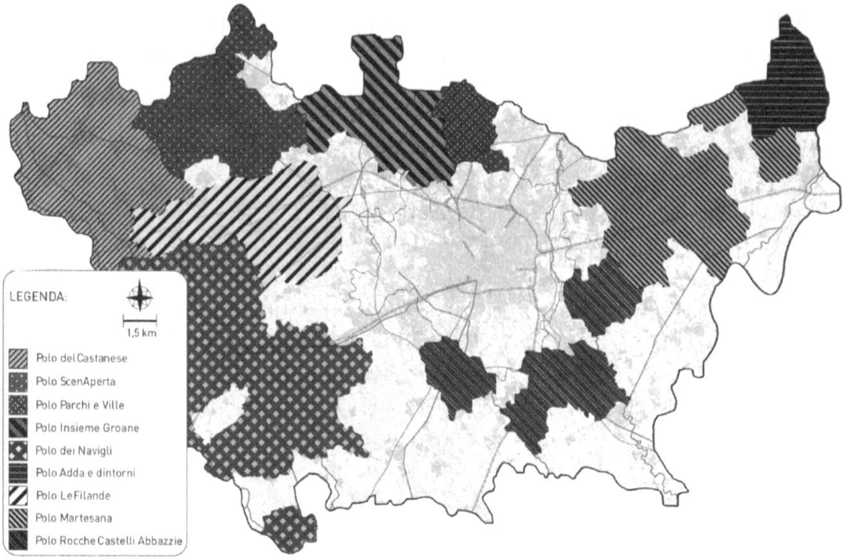

Figure 7.1 Map showing the clusters of municipalities supported by the programme Metropòli
Source of base map: www.provincia.mi.it (accessed in April 2013), redesigned by Zachary Jones.

cooperation among these groups of SMCs also dropped because of the restriction in the public finances available in the current phase of crisis (Ponzini, Colleoni, Friel, Re, and Santagata 2014).

The Northern Milanese cultural policy facing urban regeneration effects[1]

This chapter focuses on a case of joint cultural programming by seven municipalities that compose the so-called Northern Milanese subregion belonging to the first belt of Milanese hinterland (see Table 7.1). The aggregate of the Municipalities of Bresso, Cinisello Balsamo, Cologno Monzese, Cormano, Cusano Milanino, Paderno Dugnano, and Sesto San Giovanni include a complex territory of about 310,000 inhabitants and sixty square kilometers. One of the structuring elements in this territory is derived from twentieth-century industrialization and urbanization; both the economic base and the urban landscape owe much to the industrial growth. Most of these seven cities were directly impacted by the industrial plants (notably Bresso, Cinisello Balsamo, Paderno Dugnano, and Sesto San Giovanni) or by hosting the rising working-class population (e.g. Cologno Monzese). The transition to a post-Fordist development model has been difficult but also witnessed some success in the localization of specialized, creative industries and services. During the last ten years, institutional and social networking helped the design and implementation of several inter-municipal policies, and it gave birth to a formal organization (the Northern Milan Development Agency,

Table 7.1 Demographic dimension and roles of the seven cities of the "Nord Milano" partnership at the end of the 2010s

City	*Population size (approximate)*	*Role in the "Nord Milano" network*
Bresso	25,000	Partner
Cinisello Balsamo	75,000	Main promoter
Cologno Monzese	45,000	Marginal partner
Cormano	20,000	Core partner
Cusano Milanino	20,000	Partner
Paderno Dugnano	45,000	Partner
Sesto San Giovanni	80,000	Core partner

then Milano Metropòli) and to formal programmes and plans (e.g. the Northern Milan Strategic Plan, Northern Milan Development Charter, and the programming as one Pole that was financed by the Province of Milan).

Due to a dramatic change in the production base and to the opportunities and resources provided by a significant season of urban redevelopment projects, the seven medium-sized cities in the Northern Milanese region decided to network in order to address common problems of oversupply and potential redundancy in local cultural services and the arts. The municipalities promoted a cultural network, following the positive experience of the Pole "Nord Milano." The aim was to assemble, rescale, and start the reprogramming of the governance network at the subregional level. A set of cultural services contributed to local cultural and social life in the seven municipalities: two museums and cultural centres, more than a dozen theatres working as cinemas too, a variety of educational and research centres, and ten libraries. Most of these services are supported with public funds and deal with very local conditions in terms of both programming and implementation, and they attract primarily local audiences.

In the last few years, a number of large new facilities appeared in the seven cities. Many new cultural facilities were created using resources derived from the processes of urban development and redevelopment projects connected with the regeneration of former industrial sites. The overall capital investment in the period 2004–2009 exceeded 15 million euros. For example, the Toy and Childhood Museum in Cormano was a component of the conversion of a cotton factory brownfield into a medium-income housing neighbourhood in a highly accessible area adjacent to the railway station. The specialization and accessibility of this centre was supposed to attract an audience that goes beyond those living in the immediate surroundings, but it did not fully succeed. The Area Metropolis 2.0 cinematographic centre and the new multimedia Library Spazio Tilane, both in Paderno Dugnano, were part of a larger process of transformation promoted by the local government in partnership with the developer of a large former industrial site and a nonprofit organization specialized in film operating at the regional level. The number of libraries has been growing steadily, including new facilities and projects consisting of thousands of square metres, which is the case of the new multimedia library of Cinisello Balsamo. A relatively smaller library was to be included in the early 2010s in an existing historic building in Cusano Milanino, Palazzo Omodei.

As described above, the regeneration of significant parts of this urban region through the conversion or reuse of industrial sites has often been oriented to the creation of new cultural facilities and services. At the end of the first decade of the 2000s, several municipalities realized that in some cases cultural services had a regional reach and that some of them duplicated other functions already existing in nearby territories (as shown in Figure 7.2). The creation of a cultural system with a subregional spatial basis was then promoted with the objectives of coordinating different interventions and the future management of several cultural services; inducing economies of scale, scope, and proximity; and lowering a set of fixed costs related to fundraising, communication, and other activities (DiAP 2009). The image of the system leaned on the industrial and labour cultural heritage and leveraged the new intervention in the field of multimedia.

A large consultation on the main objectives of the cultural system was started by the public administration and included a group of nonprofit and private actors in the field of cultural production. A formal organization for the Northern Milanese subregion was created in 2009, the Temporary Association of Cultural

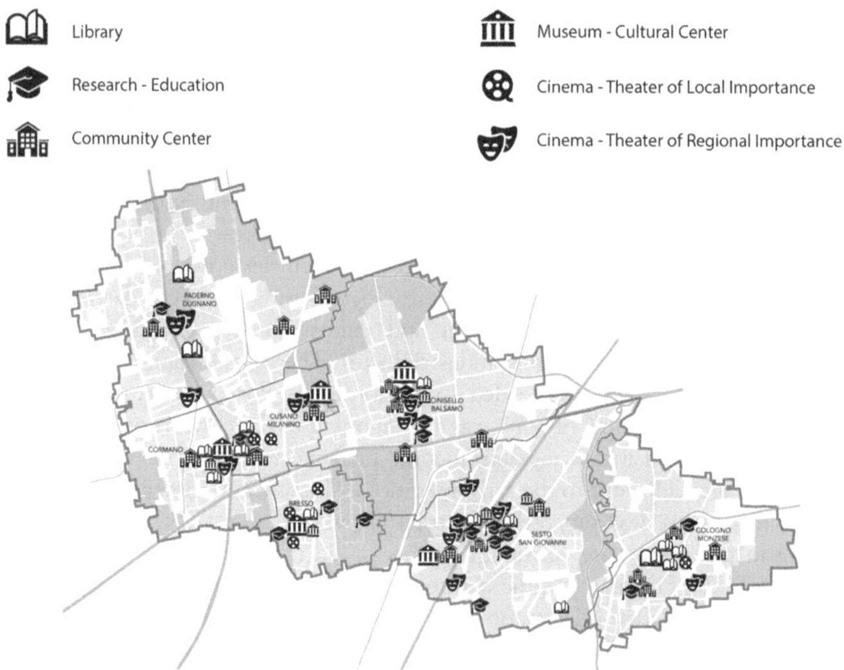

Figure 7.2 Localization of cultural facilities and services in the seven cities of the Northern Milanese region

Source of information: DiAP (2009), designed by Zachary Jones.

Services, with the participation of the seven municipalities and the provincial government. At the subregional level, another group of cultural facilities was supposed to be completed by 2010 and integrated in the new management system by the organization. The organization also aimed to propose packages that linked different cultural services with complementary economic and social activities, both inside and outside the region, targeting different groups of local users and short-stay business tourists. A number of other marketing, communication, and educational measures were conceived in order to link selected cultural activities at the subregional level.

Considering the broader urban region, many cultural projects promoted both by the Municipality of Milan and by the Lombardy Region government have been discussed. The debate includes facilities and services that will potentially be detached from or even duplicate what already exists in the region. The attempt at making the industrial sites of Sesto San Giovanni a tourist destination through a candidacy as a UNESCO World Heritage Site did not take into consideration the competition with more established attractions such as the nearby industrial plant and workers' housing settlement of Crespi d'Adda. In another instance, Renzo Piano outlined a development project in the 1,400,000 m^2 abandoned brownfields of a steel factory in Sesto San Giovanni, including a new train station, a marketplace, research and education facilities, office space, and residential towers. In addition, the municipality supported the idea of creating a new museum for contemporary art, which did not exist in the region, but was claimed (and never built) by another real estate developer in the city centre of Milan.

Despite some failures, these sorts of interventions had a high political visibility and were referred to strong political constituencies and relevant interest groups in the real estate market. Local politicians in the network apparently did not fully recognize that a broader set of economic activities would have to support the management of these cultural services in a down-phase of the real estate market and especially in such peripheral locations, compared to the strong position of the metropolitan core of Milan.

Two further examples can cast light on the current situation, when the opportunities for cooperation seem gone. The Museum for Contemporary Photography started operating in the mid-2000s in Cinisello Balsamo. Strong public support (primarily from the city of Cinisello Balsamo and the Province of Milan) and the prospect of becoming institutionalized in a stable national museum (and then gaining the support of the Ministry for Cultural Heritage) have characterized the cultural programming. But the presence of international artists, a commitment to important public archives, and advanced research could not compensate for the limited accessibility of this location. In the fall of 2013, after a period of crisis, the museum pleaded for a more central location in the city of Milan. Another former industrial site, the so-called "Carro Ponte" in Sesto San Giovanni, succeeded in becoming a centre for concerts and performing arts, serving the metropolitan region especially during the summer when it is a good place for late-hour open-air events. This allowed the whole compound to gain high exposure and to preserve its cultural activities (Ponzini, Colleoni, Friel, Re, and Santagata 2014).

During the planning phase, the activation of the subregional cultural management project encountered several criticalities. The success and activation of projects seemed to depend more on the urban regeneration and real estate appreciation dynamics and processes they were part of, rather than on the quality or actual efficiency of the proposed cultural service. In many cases, the cultural programme behind new structures was solid, but in others it did not consider basic planning factors such as the regional demand, the detailed functioning, or potential specialization in the supply of cultural services. Sometimes the accelerations or slowdowns in the regeneration processes induced problems in the coordination of cultural policy making, which, at this stage, could not be solved by the creation of a subregional management system. For example, the duplication of large libraries with copycat cultural contents and programmes will seemingly induce competition among public services for the same suburban population, rather than broadening and specializing the cultural offering.

New cultural facilities sometimes have been localized in regenerated areas or have been conceived in partnership with private parties, which are typically more interested in the final economic value rather than the social or cultural impact of their project. At the same time, the multiplication and raising of capital investments derived from real estate value capture and redistribution have not been matched by the respective operating budget for cultural programming and management or by the diffused partnering with nonprofit or private actors. In the current phase, a decrease in real estate investments and the general effect of financial restraint have generated higher uncertainty, both for the completion of new facilities and for the public and nonprofit management of some of the existing ones.

Conclusions

Since the 2000s, the creation of cultural facilities in small and medium-sized cities sometimes depended on the resources that were made available in urban regeneration processes in postindustrial brownfields and areas. Part of the increase in real estate values was effectively dedicated to the creation of new cultural facilities or the conversion of existing structures into museums, theatres, libraries, and the like. Nevertheless, cultural policy management could not generate the conditions for the small and medium-sized municipalities to sustain the cultural service provision in the long term. A number of critical aspects emerged with reference to the connection between real estate appreciation and the long-term management of cultural facilities, and the integration of different policies and the creation of networks that are capable, on a given spatial basis, of capturing and reinvesting the positive externalities of cultural production and consumption (e.g. creative production, retail- and tourism-related businesses; see Ponzini 2009; Ponzini and Rossi 2010). If compared to the core metropolitan area of Milan and, more generally, to larger Italian cities, one can see the importance of these conditions (Santagata 2014). The rescaling of such cultural initiatives is required and can grant vitality to specific places, but sometimes this depends not on the SMCs' networking activities but on more structural conditions (e.g. real estate appreciation trends and institutional investments such as those of the provincial government).

The problem of promoting cultural initiatives in a region that is dominated by the metropolitan core of Milan was addressed by assembling part of the local cultural offer in order to rescale it. The network tried to promote outstanding projects such as a national museum of photography and the creation of a new contemporary art museum serving the whole region (since the city of Milan still does not have one) in order to compete globally, but on the way it encountered significant political and financial difficulties. One specific issue can be referred to as the coupling between real estate trends and the creation of mere facilities for culture, without having a critical view on the potential competition of other, more central, cultural institutions and enterprises.

Another matter of scale is evident in this metropolitan area. The lack of a strategic vision or at least of a set of priorities for medium-term cultural programming at the regional scale makes the smaller entities suffer from the highest degree of uncertainty while facing the challenges of competing in a metropolitan cultural market. Some of these issues are now receiving more attention in local public debate, but one cannot take for granted that the negative effects and the loosening of network relationships can be easily reverted. Analysis of cultural policy making in other areas in the core of the Milanese metropolitan region showed similar criticalities regarding the coupling between real estate development projects and cultural policy making and service management in the long term (Ponzini , Colleoni, Friel, Re, and Santagata 2014).

Note

1 This case study is based on extensive quantitative and qualitative research carried out by the Department of Architecture and Planning of the Politecnico di Milano in 2009 for an agency of the provincial government, Agenzia Sviluppo Milano Metropoli (DiAP 2009). The case was partially published in Ponzini (2010a).

References

Bagnasco, A., and Le Galès, P. (2000) *Cities in Contemporary Europe*. Cambridge: Cambridge University Press.

Baker, J. R. (2006) Recruitment to boards and commissions in small cities: Individual versus contextual explanations. *State and Local Government Review*, 38(3), 142–155.

Bell, D., and Jayne, M. (2006) *Small Cities: Urban Experience Beyond the Metropolis*. London: Routledge.

Bullen, C. (2011, December) European Capital of Culture as a regional development tool? The case of Marseille-Provence 2013. *Tafter Journal*, no. 42, 8 pp.

Codecasa, G., and Ponzini, D. (2011) Public private partnership: A delusion for urban regeneration? Evidence from Italy. *European Planning Studies*, 19(4), 647–667.

Department of Architecture and Planning, Politecnico di Milano (DiAP) (2009) *Indagini finalizzate alla progettazione e attivazione di un sistema culturale integrato del Nord Milano*. Milan: Internal report conducted by Dr. Davide Ponzini.

European Commission – Directorate General for Regional Policy (2011) *Cities of Tomorrow: Challenges, Visions, Ways Forward*. Brussels: European Commission.

European Observation Network, Territorial Development and Cohesion (ESPON) (2006) *Programme ESPON 1.4.1, Small and Medium-Sized Towns (SMESTO)*. Luxemburg: ESPON.

Inteligência em Inovação – Centro de Inovação (INTELI) (2011) *Creative-Based Strategies in Small and Medium-Sized Cities: Guidelines for Local Authorities*. Lisbon: INTELI.

Kearns, A., and Paddison, R. (2000) New challenges for urban governance. *Urban Studies*, 37(5–6), 845–850.

Laurila, H. (2004) Urban governance, competition and welfare. *Urban Studies*, 41(3), 683–696.

Le Galès, P. (2002) *European Cities: Social Conflicts and Governance*. New York, Oxford: Oxford University Press.

Le Galès, P., and Therborn, G. (2010) Cities. In S. Immerfall and G. Therborn (eds.), *Handbook of European Societies* (pp. 59–89). Heidelberg: Springer.

Lorentzen, A., and van Heur, B. (eds.) (2013) *Cultural Political Economy of Small Cities*. London: Routledge.

Moulaert, F., MacCallum, D., Mehmood, A., and Hamdouch, A. (2013) *The International Handbook on Social Innovation: Collective Action, Social Learning and Transdisciplinary Research*. Cheltenham: Edward Elgar.

Moulaert, F., Martinelli, F., Swyngedouw, E., and Gonzalez, S. (2005) Towards alternative model(s) of local innovation. *Urban Studies*, 42(11), 1969–1990.

Paris, D., and Baert, T. (2011) Lille 2004 and the role of culture in the regeneration of Lille Métropole. *Town Planning Review*, 82(1), 29–44.

Petraroia, P. (2003) Il Governo. In C. Barbati, M. Cammelli, and G. Sciullo (eds.), *Il diritto dei beni culturali* (pp. 153–172). Bologna: il Mulino.

Ponzini, D. (2008) *Il territorio dei beni culturali. Interpretazioni strategiche del processo di privatizzazione dei beni e delle attività culturali in Italia*. Rome: Carocci.

Ponzini, D. (2009) Urban implications of cultural policy networks: The case of the Mount Vernon Cultural District in Baltimore. *Environment and Planning C: Government and Policy*, 27(3), 433–450.

Ponzini, D. (2010a) Cultural policymaking in the face of urban regeneration: The case of the Northern Milanese subregion. In A. Ince (ed.), *Cultural Policy and Management Yearbook 2010* (pp. 79–88). Istanbul: Istanbul Bilgi University Press.

Ponzini, D. (2010b) The process of privatization of cultural heritage and the arts in Italy: Analysis and perspectives. *International Journal of Heritage Studies*, 16(6), 508–521.

Ponzini, D., Colleoni, F., Friel, M. M., Re, A., and Santagata, W. (2014) *Progetti culturali e sviluppo urbano. Visioni, criticità e opportunità per nuove politiche nell'area metropolitana di Milano*. Milan: Quaderno dell'Osservatorio della Fondazione Cariplo.

Ponzini, D., Gugu, S., and Oppio, A. (2014) Is the concept of the cultural district appropriate for both analysis and policymaking? Two cases in Northern Italy. *City, Culture and Society*, 5(1), 75–85.

Ponzini, D., and Rossi, U. (2010) Becoming a creative city: The entrepreneurial mayor, network politics and the promise of an urban renaissance. *Urban Studies*, 47(5), 1037–1057.

Ponzini, D., and Ruoppila S. (2011, December) What's the "city" in the design and implementation of the European Capital of Culture? An open issue. *Tafter Journal*, no. 42, 7 pp.

Santagata, W. (2014) *Il governo della Cultura: Promuovere sviluppo e qualità sociale*. Bologna: il Mulino.

Servillo, L., Atkinson, R., Smith, I., Russo, A., Sykora, L., and Demaziere, C. (2012) *TOWN: Small and Medium Sized Towns in Their Functional Territorial Context – Inception report*. Luxemburg: European Observation Network, Territorial Development and Cohesion (ESPON).

Soja, E. W. (2000) *Postmetropolis: Critical Studies of Cities and Regions*. Oxford: Blackwell.

Talone, A. (2011, December) Pécs2010: Borderless culture? *Tafter Journal*, 42, 7 pp.

8 European Capitals of Culture and urban diplomacy

Rolf Hugoson

The long-term effects of the European Capitals of Culture (ECoC) project are difficult to foresee, even if individual cities have been studied and evaluated (e.g. Garcia and Cox 2013). Yet the only apparently ephemeral ECoC project might improve forms of urban cultural diplomacy. Already, plans to compete for nominations to become a cultural capital will create opportunities for negotiations – with other cities and with higher authorities (in nation-states and in the European Union) as well as with various cultural networks. Sustainability is not a static concept, since the world is made up of flows. Sustainability is created when cities negotiate their positions in a world of flows. To this end, cities must employ diplomacy. This has been known for a long time in national capitals. Now small and medium-sized cities are also increasingly engaged in new forms of diplomacy, as demonstrated by their efforts to contribute to the ECoC project.

Cities, flows, and networks

Residents who govern a place can be trapped within larger forces, leading to population fluctuations that might appear mystical to the residents themselves. Consider the small town Bardou in Southern France, in retrospect observed by Jane Jacobs: "Time and again, like a toy on a string, Bardou has been jerked by some external economic energy or other" (Jacobs 1984, 34). Jacobs thus observed an interplay between the place and more distant networks. But to remain a passive object is unsustainable.

Large and apparently successful cities are also acted upon by outside flows. For sure, the large American cities of the 1920s – Chicago, Detroit, Los Angeles, and New York – appeared to be triumphant: "electric with opportunity and excitement"; but, already, "powerful social and economic forces were combining to create an irresistible movement towards decentralization" (Fishman 1995 [1990], 395).

Large cities to some appeared unsustainable. Although today they have regained in importance, this is hardly because they have become autonomous. On the contrary, they depend even more upon fluid relationships: "Cities . . . are relationally constituted, a space where multiple geographies of composition intersect, bringing distant worlds into the centre of urban being and projecting the placed outwards through myriad networks" (Amin 2012, 63).

Some aspects of the interdependency between global forces and large cities are captured by John Friedmann's World City hypothesis, coined in the 1960s and later reformulated: "The form and extent of a city's integration with the world economy, and the function assigned to the city in the new spatial division of labour, will be decisive for any structural changes occurring within it" (Friedmann 1995 [1986], 69f.).

Both Jacobs and Friedmann observed global forces, but they also argued that cities could be navigated. Thus, Pierre has concluded more recently:

> Few cities escape the pressures of globalization, and the powers of a deregulated international capitalist economy is far beyond that of any city, but the process of defining "offensive" responses to those pressures does offer a choice in terms of strategy and objectives.
>
> (Pierre 2011, 134)

But where shall the mobilization take place? Some authors argue that contemporary growth-oriented and entrepreneurial approaches to urban governance are "oriented toward a fundamental rescaling of national state space and, more generally of national state sovereignty" (Brenner 2011, 152). So perhaps cities are becoming more important. Still, unequal success among cities or the logical conflict between competition and cohesion can lead to new forms of conflict (Ache et al. 2008).

Smith has argued that we must move beyond macroeconomics and do more research on "the cultural, social and political networks that bind cities together . . . how the mobilization of materials, and the practices of people, stitch together the networks between cities" (Smith 2003, 29). Agreeing on this point, I would argue that success is a result not merely of economic adaptation but also of cultural management and diplomatic strategies, matters that are combined in the European Capitals of Culture programme.

The cultural interests of cities

When thinking about cities and global flows, it might appear as a drastic reduction to focus just on one subject – culture mainly in terms of the arts, and notably the effort of some cities to become a European Capital of Culture. Let us begin by confronting the idea that this is indeed a marginal topic, something almost superfluous. Why would cities be interested in the arts?

During the nineteenth century and most of the twentieth century, it appeared more attractive for most European cities to promote industrialization because of the rapid wealth that could be produced by new technologies. In the early twenty-first century, industry is still a positive term. Even if heavy industry has lost ground in cities, manufacturing certainly remains a key interest, and housing and real estate have remained central, together with transport and communications infrastructures: from harbours and railway stations to wireless Internet. Yet, in relative terms, service sectors such as administration, education, health, and food services appear increasingly influential.

Access to supplies, to markets, and to skilled labour remain decisive for companies considering new locations. More specifically, to manage markets for goods, services, and persons, various kinds of cultural expertise and knowledge are needed. According to economists, culture thus forms part of a market where attractive goods and services are produced and where both consumption and production have grown significantly the last decades.

In a recent explorative study on behalf of the European Expert Network on Culture, the Italian economist Pier Luigi Sacco presented a model where many countries and, notably, cities are about to enter a new phase in the production and consumption of culture. In this study, Sacco outlined a history of cultural markets, from the Renaissance through the welfare state to the present, from "Culture 1.0" to "Culture 3.0" (Sacco 2011).

The first phase was concerned with highly expensive goods and services produced for a small elite,

> people with large financial possibilities and high social status, who derived their wealth and status from sources other than cultural commissioning in itself, but decided to employ some of their resources to ensure that cultural producers could make a living, thereby getting the possibility to enjoy the outcome of creative production and to share it with their acquaintances.
>
> (Sacco 2011, 2)

In the next phase, "Culture 2.0," the dispersion of wealth in society together with new technologies such as radio and television made cultural goods and services more accessible to wider audiences:

> [A]udiences expand significantly, whereas cultural production is still severely controlled by entrance barriers as the access to productive technologies is difficult and financially expensive, so that would-be cultural producers are filtered by complex selection systems, that differ from one cultural sector to another.
>
> (Sacco 2011, 3)

Finally, we might be entering "Culture 3.0," a phase where potentially everybody becomes a cultural producer, when we gain access to new interactive technologies:

> innovations that, unlike the previous one, not only cause an expansion of the demand possibilities, but also, and mainly, an expansion of the production ones. Today, one can easily have access to production technology that allow professional treatment of text, still and moving images, sound, and multimedia with impressively quick learning curves and at very cheap prices . . . it becomes increasingly difficult to distinguish between cultural producers and users.
>
> (Sacco 2011, 3)

Perhaps due to the balance between elite arts and popular culture in each city, efforts to promote arts and culture will vary. Not all cities will try to enhance their roles as cultural centres. But if such engagement occurs, we should expect it to be directed outwards, towards wider flows, not just organised internally as municipal cultural policy.

To the extent that European cities tend to have "foreign policies," these are typically employed on two fields: first, within city-regions (that is to say, among their neighbours); and, second, by signing contracts with specific cities to create partnerships for general or specific purposes, such as regional transport policies (van der Heiden 2010).

As an example of how this could work in terms of cultural policy, we can note that when Copenhagen was a European Capital of Culture back in 1996, it cooperated with several "metropolitan counties" with the aim to market a regional economic development policy (Engelstoft and Jørgensen 1997, 233). Varieties of regional strategies were even more decisive when Luxembourg was ECoC for the second time in 2007 and when Essen in 2010 dispersed the title to fifty-two towns and cities in the whole Ruhr region.

Partnerships with more distant cities can also be established. Lausanne together with, among others, Geneva and Lyon have been participating in a cultural policy network called "Les rencontres": "Thanks to the contacts established in this network, Geneva officially supported Lyon in the competition to become the capital of culture in 2013" (van der Heiden 2010, 52). Of course, this is not a guarantee for becoming a "capital of culture": for 2013, Marseille was chosen instead of Lyon. Still, Lyon remains interested in promoting its cultural agenda.

European Capitals of Culture: top-down and bottom-up

The European Capitals of Culture project, or ECoC for short, has changed character since its invention in the 1980s. Yet, from the beginning, it has presented an interesting combination between international or even transnational integration and reinforced local government.

In the early 1980s, Europe appeared slow and hesitant on all fronts. The eight Ministers of Culture of the EEC had their very first formal meeting in 1984, although there was also a famous "first informal meeting," arranged in September 1982 by French Socialist Jack Lang and his Christian Democratic colleague Vincenzo Scotti from Italy. The result of 1984 was meagre: the ministers decided only that they should meet again.

Then arrived Melina Mercouri and Jack Lang's joint idea for a "Cultural City" event organised nationally, but with a European touch. Lang from France and Mercouri from Greece both belonged to governments made up of socialist parties. They had learned that high ambitions had to be rescaled to economic restrictions: the field of culture was an ideal area for symbolic interventions that cost little but created much visibility. So in 1985 the EEC Ministers decided that they should arrange in common an annual event called "the European City of Culture" (Littoz-Monnet 2007, 48; McMahon 1995, 136).

Somewhat unexpectedly, a new international agreement was also reached in these years. French President François Mitterrand together with UK Prime Minister Margaret Thatcher and West German Chancellor Helmut Kohl decided that until 1992, the European Economic Community should develop into a single market and then become a union. In 1987, the European Commission headed by former French Minister of Finance Jacques Delors also argued that culture was important: "[I]ncreased cultural activity is now a political as well as a social and economic necessity, given the twin goals of completing the internal market by 1992 and progressing from a People's Europe to European Union" (quoted in Littoz-Monnet 2007, 54).

The new urban European cultural policy was portrayed as coming from below rather than being dictated from the top. The aim was to bring forth the multiplicity of European cultures. To institute new forms of cooperation in terms of laws and economic arrangements, the EU had to proclaim diversity, although the relevant motto "unity in diversity" was chosen only in the year 2000 (Fornäs 2012). The ECoC project in particular was connected to a vision where culture appeared: "more as a project and a co-operative construction rather than in terms of inheritance of and belonging to fixed cultural contents" (Sassatelli 2008, 241).

A new engagement for cities in Europe, carried not just by traditional local elites and property owners but also by new social movements, was fertile ground for Lang and Mercouri's new project. For example, a famous Italian project had already been invented by architect Renato Nicolini in 1977 when the streets of Rome were reclaimed for a series of summer festivities called the "Estate Romana."

However, the new "European cities of culture" were selected by national governments. So when a right-wing mayor was elected in Rome, Italy's socialist Prime Minister Craxi instead nominated Florence as Italy's "Cultural Capital" for 1986. In Florence, fellow socialist Massimo Bogianckino had just been elected, returning from a job as the Intendant of the Paris Opera. Mercouri had already selected Athens for 1985. The Dutch chose Amsterdam for 1987, formally the Dutch capital, despite the Hague's role as seat of government. West Germany's government also preferred an alternative to its small capital Bonn, and it opted for West Berlin, in 1988 situated in the middle of East Germany.

The ECoC project was thus hardly the result of fully coordinated European policy. Rather, the project was an effect of the French willingness to move ahead with what has been termed "cultural cooperation *à la carte*." In an early "blue book," the French Government explained: "[R]ather than exhausting ourselves trying to reach an agreement between twelve states on the principles and means of an EC measure, we will try to take swift action together with those member states who wish to be involved" (quoted in Littoz-Monnet 2007, 51).

Patel (2013) recently coined the term "integration through interpellation" for fields where only those who listen to the European calls participate in the official transnational integration processes. In the long run, the ECoC project's intergovernmental and thus partly decentralized character increasingly opened up the possibility of cities other than the national capitals being elected "capitals of culture." Indeed, the term "capital of culture" became quasi-official already through the selection of Florence in 1986 and also helped promote formal "non-capitals" such as Amsterdam, Berlin, and Glasgow in 1990; Antwerp in 1993; Thessaloniki in 1997; and Weimar in 1999. It became increasingly clear that the ECoC project offered cities a new transnational diplomatic arena.

What is urban diplomacy?

Originally, the term *diplomacy* had more to do with keeping an archive than with the management of relationships with other states. The modern use of the term *diplomacy* was introduced in English only in 1796. But, of course, established practices for negotiations between states had then already evolved for centuries. Indeed, these negotiating practices began to take modern shape already during the thirteenth and fourteenth centuries, *among cities*:

> The Italian city states . . . were interconnected by countless common interests as well as sundered by ferocious rivalries; they were constantly engaged in a competition for power and preoccupied by those combinations and alliances which might render that power predominant.
>
> (Nicolson 1939, 29f.)

In this short description we encounter some basic ingredients that allow for diplomacy to occur as negotiation between cities. First, there must be a plurality of agents with some amount of autonomy. Second, the competition must not be so fierce as to shadow the opportunities for cooperation – there should be both common and separate interests. Third, even when competition becomes fierce, the need for cooperation tends to reinstate itself when new alliances are formed. Fourth, the diplomat will be the agent who has the capacity to understand the game, to discern the threats and the opportunities.

Diplomacy long appeared as a secret club for the initiated few to remain influential. But the advent of the telegraph and mass media in the 1840s necessitated that reactions of wider audiences be taken into account in diplomatic practice. The idea of "public opinion diplomacy" was formulated during "the Springtime of peoples," the revolutionary year of 1848, by the liberal Carlo Cattaneo from Turin (Hugoson 2004).

As we know, cities in most countries were dominated by the government of the nation-state, and that is why cities had to direct their "diplomatic" endeavours towards the government. Arguably, the most important arena for local lobbying on the national level remained the national parliament.

In most countries, there also exist particular organizations of cities – for example, the Local Government Association in the United Kingdom, the Spanish Federation of Municipalities and Provinces, or the Swedish Association of Local Authorities and Regions. Governments can be centralized both on the national and on the transnational level – in modern times, perhaps the European Union; in older times, various forms of empires. The EU also has a prestigious assembly for local and regional authorities called the Committee of the Regions (www.ccre.org;

http://cor.europa.eu). Beyond official authorities, local agents also network in institutions such as chambers of commerce (see Benoit 1995).

As observed already by Nicholson, in diplomatic practice we must include a variety of efforts to influence and persuade other agents: not just governments, but also peoples; not just by means of formal protocols, but also with the help of new information technologies and, of course, through informal and personal networks of contacts. In general, diplomacy is a more flexible practice than mere membership in formal organizations. Diplomatic activities occur on quite different levels of authority. Their character will change depending upon the identity of their agents: politicians, civil servants, NGOs, companies, or sometimes ordinary citizens. This variety of practices is thus more decisive than our choice of a particular term to focus the subject. We might for example also talk about urban or municipal paradiplomacy, urban foreign policy, transnational networks of cities, or city diplomacy.

Early examples of urban diplomacy through cultural investments were the great industrial exhibitions known as World Fairs. These events were often accompanied by older forms of popular entertainment, as Greenhalgh (1988) observed: "Sport, fairground stall and rides, musical and comic troupes, sellers of food and ephemera all derived originally from the annual market-fairs held around Europe for centuries" (45).

Even more prestigious as a form of diplomacy were the Olympic Games of the ancient Greek city-states, which were combined with religious festivities in cities where theatre and drama had central importance. However, the links between antique and modern forms of cultural festivities are far from continuous. When the Olympic Games and international exhibitions returned to the agenda in the late nineteenth century, they were without exception sponsored by nation-states. The Olympic Games typically enhanced nationalism and state sovereignty. In other forms of cultural diplomacy, it was the nation-state that sponsored national cultures as well as prestige institutions in the national capitals. Yet, through these events, new international stages and arenas were also constructed – typically in cities (see Roche 2000).

ECoC as a new form of urban cultural diplomacy

In 1995, when preparing for the ECoC year 2000, the Ministers of Culture fell for the temptation to allow the list to be inflated. The cities of culture were declared to be Avignon, Bergen, Bologna, Brussels, Helsinki, Krakow, Prague, Reykjavik, and Santiago de Compostela, and it has been reported that the candidates were shocked by being lumped together in such numbers (Gold and Gold 2005).

The nine cities of the year 2000 were in 1995 required by the Council of Ministers "to co-ordinate their programme, define a common theme for this event and act together in the organisation of a European cultural space for the year" (quoted in Cogliandro 2001, 45). In response to this, the nine cities established a formal network and a new legal entity: Association of the Nine European Cities of Culture of the Year 2000 (AECC). Agreement for the creation of AECC was reached in 1996, and it was formally established in 1998. The nine cities would "work to make the culture of every city and country known to each other and to all the people of Europe and the whole world" (Cogliandro 2001, 74f.).

The AECC was accompanied by relevant administrative procedures, such as the choice of two working languages, English and French; the establishment of a board with mayors or cultural directors; and a rotating presidency. The cities also went ahead with a common marketing project and a number of coordinated projects. However, the marketing efforts were not impressive. For example, the AECC failed to establish a common website, managing only to create links between the websites of each city, and the common sponsorship campaign was a complete failure (Cogliandro 2001; Gold and Gold 2005).

In preparation for the year 2000, each city also coordinated at least one cultural project intended to promote the nine cities as a community. Some of these projects were rudimentary, but others were more ambitious and deemed successful. For example, Helsinki and Bologna arranged a project called CaféNine. A café in each city was selected and equipped with a screen, which functioned as a kind of window into one or several of the other eight cafés (Cogliandro 2001).

Beginning in 2001, the number of capitals decreased, but two cultural cities were still promoted every year. First, Rotterdam and Porto; then, Bruges and Salamanca; in 2003, Graz on its own; then, Genoa and Lille. Furthermore, in 1999, the list of countries for the years 2005 to 2012 was fixed. A new designation process now began, although it took some time before it was fully implemented. One early result was that each country had more time to arrange a competition. Assessments of candidate cities by an international group of experts became increasingly important.

The European Parliament and the Council of the European Union had decided to call the European Capital of Culture scheme a "community action." The top-down perspective remained, but the effort to establish trans-nationality rather than nationality was emphasized. Beginning with Cork in 2005 and Patras in 2006, the nomination was also officially for "European Capitals of Culture." Sibiu and Luxembourg followed in 2007, and in 2008 Liverpool and Stavanger were elected capitals of culture. In Ireland, the government promoted Cork, Galway, Limerick, and Waterford, before allowing an international panel to select Cork for 2005. In preparation for 2008, the United Kingdom had allowed twelve cities to compete. The government followed the international panel's recommendation of Liverpool (Griffith 2006).

It was decided that from 2009 the programme would promote exchange between older member states in the West and newer members in Eastern Europe. The first new East-West couples were Vilnius and Linz in 2009; in 2010, Pecs and Essen – and also Istanbul; in 2011, Tallinn and Turku; in 2012, Maribor and Guimarães. In 2013, the capitals were Marseille and Kosice; in 2014, Umeå and Riga; in 2015, Mons and Plzen; in 2016, Donostia-San Sebastian and Wroclaw; in 2017, Aarhus and Paphos; in 2018, Valletta and Leeuwarden; and in 2019, Plovdiv and Matera (http://ec.europa.eu/culture).

Since the 2008 selection of the 2013 capitals, an international committee has formally led the process. The general tendency is that national governments' choices to promote specific cities or regions are decreasing. Yet, as indicated above, in terms of European Union law, the subsidiarity principle remains particularly vigorous on the fields of cultural policy, declared to be a national domain.

The increasing recourse to competitions has increased the number of cities interested in becoming "capitals of culture." The selection clearly depends upon the experience and the engagement of particular cities. Competition between cities rather than cooperation seems most characteristic. After all, there can be only two capitals each year: the nine of the year 2000 or the three of 2010 are not recalled as successes. Still, a number of cities had the ambition to become European Capitals of Culture in 2013 and in 2014, even if they were later discarded in favour of Marseille, Kosice, Umeå, and Riga.

The culturally ambitious cities from the four countries concerned were: Amiens, Banská Bystrica, Bordeaux, Bratislava, Cēsis, Dolný Kubin, Gävle, Jūrmala, Lund, Lyon, Martin, Nice, Nitra, Norrköping, Prešov, Saint-Etienne, Strasbourg, Toulouse, Trenčín, Trnava, and Uppsala. This impressive list can be added to further if we consider the less successful candidacies for 2015 and 2016, which included Alc. De Henares, Asturias, Bialystok, Caceres, Cuenca, Gdansk, Katowice, Liepaja, Lodz, Lublin, Malaga, Murcia, Oviedo, Pamplona, Palma de Mallorca, Poznan, Santander, Szczecin, Tarragona, Tenerife, Torun, Valencia, Warsaw, and Wroclaw; and Burgos, Cordoba, Las Palmas de Gran Canaria, Ostrava, Segovia, and Zaragoza (www.ecoc-doc-athens.eu).

On their own, such lists of interested cities do not constitute proof that cultural diplomacy is becoming more important, but they certainly indicate a possibility. Each of these candidate cities, whether successful or not, has promoted and to some extent tried to change its own identity as a "cultural city." Simultaneously, each city has contributed to a small change in the idea of what a European Capital of Culture can be. Finally, the candidate cities have tried to internalize (formally or with more engagement) notions about cities and culture promoted by the European Union and its experts.

There is room for conflicts and alliances between cities, and the project can be assumed to establish three kinds of diplomatic relationships:

(a) when the candidates compete, there will be *strategic efforts* to appear cultural, while similar efforts of other cities will be monitored;
(b) *cooperative efforts* occur when policy learning is arranged between former, present, and upcoming cities; and
(c) after cities have competed and cooperated in the ECoC project, they might employ *transformative efforts* and use their investments in social capital to organize similar projects, such as festivals or sports event.

The strategic cultural ambitions of cities evidently need not be limited to becoming ECoCs. Indeed, as indicated above in the overview of Sacco's Culture 3.0 study, new forms of culture tend to impregnate ever more scenes of production and consumption. It might be the case that, at the moment, "culture" and "creativity" are mentioned more often together with more or less traditional forms of city planning, as municipal authorities together with estate developers or residents try to make their cities or their neighbourhoods more attractive.

Some of these rather limited forms of cultural strategies are probably enforced by cities monitoring each other. Some cities will appear as exemplars to be followed: early on, Glasgow managed to portray its ECoC year as a success story, Barcelona has been a famous icon of modernity since the late nineteenth century, and Bilbao's effort at urban renewal became famous in 1996 through the new museum designed by Frank Gehry. These latter two celebrity cities have been studied by Sara González, who focused on networks called "policy tourism," where delegations from cities all over the world come to Spain trying to learn from the urban experiences of Barcelona and Bilbao (González 2010).

Of course, as scholars, we might find it equally interesting to study cities that, according to reputation, failed to deliver a successful or memorable ECoC year, such as Paris in 1989. In any case, beginning from the monitoring of others, more focused efforts at cooperation can perhaps follow. Even if many cities experience the same flows of capital and ideas, they will experience them differently depending upon their diverse mixtures of local agents from various sectors as well as varying presences of stability, tolerance, and participatory practices. Thus, for a visiting delegation to actually experience another city (rather than merely learning from a distance) contributes to a more fundamental change of perspectives. Such visits can also provide a learning process for the host (McCann 2011).

To this end, the European Commission has encouraged cooperation between previous, present, and future ECoC cities, who decide for themselves for how long after their ECoC experience they want to cooperate, and for how long in advance. Thus, at a "family meeting" in Sweden, a year before Umeå's ECoC year, sixteen other past, present, and future capitals were present: Liverpool 2008, Stavanger 2008, Linz 2009, Ruhr 2010, Tallinn 2011, Turku 2011, Guimarães 2012, Maribor 2012, Kosice 2013, Marseille 2013, Riga 2014, Mons 2015, San Sebastian 2016, Wroclaw 2016, Århus 2017, and Valletta 2018 (www.umea2014.se).

Such family events appear rather formal, but just like the Capital year itself, they can also be arenas where various kinds of practitioners meet and engage in more focused cooperative activities. An example of cooperation along these lines of meetings between practitioners and theorists is the academic conference organised yearly in one of the ECoC university cities by the "University Network of European Capitals of Culture," or UneECC (www.uneecc.org).

Again, what matters are not just the formal arenas but rather the apparent increase in diplomatic experience. Complexity in general is typical of any city, but the empirical particularities will vary, and thus different cities are differently prone to manage their cultural capital experience by reinvesting in new projects. But, as indicated above, beyond this rather special transnational European project, there is a whole range of recurring events that cities might bid for, not just cultural festivals, but also European Capital of Sports, European Capital of Youth,

Olympic Games, Commonwealth Games, national championships in various sports, or titles connected to anything from ecological sustainability to infrastructure management. For example, the European Commission in 2010 instituted the European Green Capital award (http://ec.europa.eu/environment). The more an event manages to engage different groups within and beyond the city, the more opportunities there will be for people to participate in its continuous transformation to a more sustainable future.

References

Ache, P., Andersen, H. T., Maloutas, T., Raco, M., and Tasan-Kok, T. (eds.) (2008) *Cities Between Competitiveness and Cohesion: Discourses, Realities and Implementation*. Heidelberg: Springer.

Amin, A. (2012) *Land of Strangers*. Oxford: Polity Press.

Benoit, S. (1995) Local policies to attract mobile investment. In P. Cheshire and I. Gordon (eds.), *Territorial Competition in an Integrating Europe* (pp. 222–243). Avebury: Aldershot.

Brenner, N. (2011) Urban locational policies and the geographies of post-Keynesian statehood in Western Europe. In D. Davis and N. Libertun de Darun (eds.), *Cities and Sovereignty: Identity Politics in Urban Spaces* (pp. 152–175). Bloomington: Indiana University Press.

Cogliandro, G. (2001) *European Cities of Culture for the Year 2000: Final Report*. Athens: European Commission and Association of the European Cities of Culture of the Year 2000 (AECC/AVEC).

Engelstoft, S., and Jørgensen, J. (1997) Copenhagen: A redistributive city? In C. Jensen-Butler, A. Sachar, and J. van Weesep (eds.), *European Cities in Competition* (pp. 209–243). Aldershot: Avebury.

Fishman, R. (1995) [1990] Megalopolis unbound (repr.). In P. Kasinitz (ed.), *Metropolis: Centre and Symbol of Our Times* (pp. 395–417). London: Macmillan.

Fornäs, J. (2012) *Signifying Europe*. Bristol: Intellect.

Friedmann, J. (1995) [1986] The World City hypothesis (repr.). In P. Knox and P. Taylor (eds.), *World Cities in a World System* (pp. 317–331). Cambridge: Cambridge University Press.

Garcia, B., and Cox, T. (2013) *European Capitals of Culture: Success Strategies and Long-Term Effects*. Brussels: European Parliament, Directorate General for Internal Policies.

Gold, J., and Gold, M. (2005) *Cities of Culture: Staging International Festivals and the Urban Agenda, 1851–2000*. Aldershot: Ashgate.

González, S. (2010) Bilbao and Barcelona 'in motion': How urban regeneration 'models' travel and mutate in global flows of policy tourism. *Urban Studies*, 48(7), 1397–1418.

Greenhalgh, P. (1988) *Ephemeral Vistas: The* Expositions universelles, *Great Exhibitions and World's Fairs, 1851–1939*. Manchester: Manchester University Press.

Griffith, R. (2006) City/culture discourses: Evidence from the competition to select the European capital of culture 2008. *European Planning Studies*, 14(4), 415–430.

Hugoson, R. (2004) 1848: Den internationella politikens enastående år, speglat i samtida italiensk press. *Historisk tidskrift* (Sweden), 123(1), 3–22.

Jacobs, J. (1984) *Cities and the Wealth of Nations: Principles of Economic Life*. Viking: New York.

Littoz-Monnet, A. (2007) *The European Union and Culture: Between Economic Regulation and Cultural Policy*. Manchester: Manchester University Press.

McCann, E. (2011) Urban policy mobilities and global circuits of knowledge: Toward a research agenda. *Annals of the Association of American Geographers*, 101(1), 107–130.

McMahon, J. (1995) *Education and Culture in European Community Law*. London: Athlone.

Nicolson, H. (1939) *Diplomacy*. London: Thornton Butterworth.

Patel, K. (2013) Integration by interpellation: The European Capitals of Culture and the role of experts in European Union cultural policies. *Journal of Common Market Studies*, 51(3), 538–554.

Pierre, J. (2011) *The Politics of Urban Governance*. London: Palgrave Macmillan.

Roche, M. (2000) *Mega-Events Modernity: Olympics and Expos in the Growth of Global Culture*. London: Routledge.

Sacco, P.-L. (2011) Culture 3.0 – a new perspective for the EU 2014–2020 structural funds programming. Available at: www.eenc.info/eencdocs/papers-2/culture-3-0-%E2%80%93-a-new-perspective-for-the-eu-2014-2020-structural-funds-programming/ [Accessed 12 December 2014].

Sassatelli, M. (2008) European cultural space in the European Cities of Culture: Europeanization and cultural policy. *European Societies*, 10(2), 225–245.

Smith, R. G. (2003) World city actor-networks. *Progress in Human Geography*, 27(1), 25–44.

van der Heiden, N. (2010) *Urban Foreign Policy and Domestic Dilemmas*. Essex: European Consortium for Political Research (ECPR).

Section III

Culture for sustainable development in urban policies and practices

9 A place in the city

Recognizing creative inclusion[1]

François Matarasso

> The state consists not merely of a plurality of men, but of different kinds of men; you cannot make a state out of men who are all alike.
>
> — Aristotle, *The Politics*, II.ii (1992, 104)

A few years ago, in the basement room of an asylum seeker hostel in Utrecht, I watched a performance by five young men from the Middle East, Africa, and Southeast Asia. The show used physical theatre, dance, music, and poetry to weave tough personal experiences with fantasies of Western cultural icons such as J-Lo and Beyoncé. It also challenged attitudes towards refugees and asylum seekers:

> Things are not always what they seem at first glance. . . .
> But people are lazy.

The young men made their work with the support of professional artists from Dox, a theatre company based in Utrecht since 1997.[2] Dox offers one-year training courses, full-time or part-time, to young people with diverse cultural and social backgrounds who work with directors, artists, and choreographers to produce new shows that are toured in the Netherlands and abroad. The young trainees also learn how to pass on their skills through peer education. *Het is maar . . .* was part of that programme and toured to Dutch festivals and refugee centres between 2005 and 2006; I saw it in the context of a community arts festival organised in 2006 by Vrede van Utrecht.

Het is maar . . . had artistic integrity, style, and panache, and it gave a platform for self-representation to a group of marginalized people who are often described with hostility by others. Indeed, it could be argued that the social marginalization of asylum seekers and other "unpopular" people is facilitated by restricting their opportunities for unmediated representation in cultural and political discourse, whose antagonism thereby goes unchallenged. But this piece also raises the question of why these young creative people are marginalized in the first place. Now that cities vie with one another to be seen as creative, contemporary, and competitive, even recognising the "diversity advantage" (Wood, Landry, and

Bloomfield 2006, 60), they might be expected to capitalise on the full potential of their talent bases. These young men would agree with Richard Florida (2008) about the importance of where you choose to live, although, unlike the secure and well-resourced "creative class" that cities are now keen to attract, these men had few obvious advantages. Their assets were human: the courage, resourcefulness, and resilience they had shown in getting to Utrecht, and the creativity, imagination, and energy they were putting into their theatre, even as they waited to find out whether they would be allowed to stay and contribute to the city they had asked to join.

There is a tension here that is only made more acute in the changing conditions of the 2007–2008 financial crisis and the prolonged recession it has caused. In the 1990s, as neoliberal economic policies became orthodoxy, the idea took hold that establishing a certain kind of creative and cultural milieu – or perhaps just an image of it – was essential to cities' capacity to attract and keep young talent (Florida 2002). Its champions often focused on a narrow section of any urban community: the affluent (or prospectively affluent) and well educated who were supposed to drive consumption, growth, and prosperity in a cultural version of trickle-down economics.[3] But in making cities more attractive to the "creative class," cultural policy risks neglecting the needs and interests of other less obviously or commercially creative groups, such as those young asylum seekers, or even increasing their exclusion from the changing urban environment. Were this to be the result, it would be unacceptable in itself and, as research increasingly suggests, undermine the value of any apparent economic benefit (Wilkinson and Pickett 2009; Stiglitz 2012). To see why, and what measures might at least mitigate some of the problematic aspects of these policies, it is worth looking again at some well-established ideas, not least the relationship between the city, culture, and creativity.

Cities are glamorous. Since the zenith of Babylon, Athens, and Rome, the city has been represented as the place where things happen, where new ideas and fashions are created, where governments are toppled, where students are educated and money made. Cities are young, even when – like Utrecht, Nottingham, or Lyon – they are ancient, because they constantly renew themselves, shedding old skins like snakes. They are concentrated and monumental, drawing in collective resources and radiating outwards (and now upwards) their economic and cultural power. Their landmarks – Notre Dame, Ponte Vecchio, or the Tivoli Gardens – signify them to the world. Cities recreate themselves symbolically with apparent ease. Few people outside Spain knew much of Bilbao before Frank Gehry endowed it with that icon of contemporary consumer tourism, the Guggenheim Museum. Millions who have never been to Bilbao now have an instant image of the city, though that may be quite unrepresentative of its citizens' daily realities.

By contrast, rural districts make do with being "not-cities," as they have since at least the time of Virgil and Horace. They are cast as territories of retreat, idealized landscapes of permanence contrasted with the city's mutability. More prosaically, rural areas offer most people only the hard, grinding labour of producing food. It has never been surprising if their brightest and hungriest youth follow the yellow brick road to the city as soon as they can. Populations grow, and the land cannot always support so many, while advances in agricultural technology reduce the need for hands. It happened in Europe in the nineteenth century, as millions left to work in cotton mills, iron works, and potteries, swelling the population of Bradford from six thousand to almost two hundred thousand between 1800 and 1850. It is happening today in China, Brazil, India, Nigeria, Mexico, and many other countries. But the *favelas* of Rio are not as photogenic as Copacabana: Oz wasn't called the Emerald City for nothing. In 2008, humanity passed a historic landmark: for the first time, most humans lived in cities. Some of them are bigger than countries: there are 145 nations with fewer citizens than any one of the world's twenty-three most populous cities,[4] and, of those megacities, only six are in the developed world (Tokyo, New York, Los Angeles, Moscow, Osaka, and Paris) – small wonder that urbanization has become such an urgent concern of policy makers, planners, and professors.

Cities are crucibles of innovation, albeit to varying degrees, so it is not surprising that creative industries, culture, and creative thinking about urban planning and management should become prominent in debates about the modern city. Charles Landry and Franco Bianchini outlined an early manifesto in 1995, arguing:

> Future competition between nations, cities and enterprises looks set to be based less on natural resources, location or past reputation and more on the ability to develop attractive images and symbols and project these effectively. Indeed the urban renewal process can itself become a spectacle, as, in David Harvey's words, aesthetics comes to replace ethics in contemporary urban planning.
>
> (Landry and Bianchini 1995, 12)

The creative city theories that emerged during the 1990s have become influential in some very different places, despite, or perhaps because of, the wide variety of approaches taken, from the narrowly instrumental to the holistic and visionary (Landry 2000; Florida 2002). Their dissemination and adoption was facilitated both by a globalising economy and by the growing importance of culture as a source of consumer goods and as a creator of meanings at a time of weakening political and religious values, at least in Europe.

The rise of the creative city parallels the rise of the neoliberal globalisation that became the dominant economic model during the 1980s under the banner of the "Washington Consensus." Among other tenets, this ideology required governments to remove restrictions on businesses' ability to move goods, capital, and skilled workers across national borders. Among its diverse consequences was increased pressure on cities to retain or attract investment in a global marketplace (although freedom of movement was a variable concept).[5] City competitiveness became established in urban policy and rhetoric. International bodies such as the OECD promoted it, while even the smallest cities began to compare themselves, not always for very clear reasons, to unspecified competitors. Place marketing became fashionable, and quality of life, including cultural facilities, was used to promote the image of cities. In Europe, Glasgow raised eyebrows when, as European Capital of Culture in 1989, it transformed a title previously awarded to candidates like Athens and Paris into an effective rebranding strategy. The "Glasgow's Miles Better" campaign won awards and became a much-imitated model. In the cultural sector, people began talking of "world-class" galleries and theatres without clarifying how that comparison might be meaningful to the local people who attended and paid for the facilities. The rhetoric of cultural self-promotion inflated throughout the 1990s in line with the overpriced stocks traded in the twenty-four-hour markets, and, like them, the connection between what some people were prepared to pay and the use value (or artistic value) of what was being bought became increasingly tenuous.[6]

The financial meltdown in September 2008 brought economic, social, and political effects that are neither understood nor exhausted. Both consequences and responses are disputed. But the critics of economic neoliberalism have long questioned its tenets on the operation, rationality, and regulation of free markets (Chang 2010) as well as the limits and transferability of market principles (Sandel 2012). But most of all they have challenged the idea that increases in economic inequality are necessary or effective in increasing overall wealth (Stiglitz 2012). In 2008, the American sociologist Immanuel Wallerstein observed, "The distribution of income worldwide and within countries became very skewed – a massive increase in the income of the top 10% and especially of the top 1% of the world's populations, but a decline in real income of much of the rest of the world's populations" (Wallerstein 2008).

Evidence for this increase in global inequality is so abundant that advocates of the policies associated with it largely accept its existence, arguing either that it is a necessary means to the greater good of increasing overall wealth or even – in the words of the current mayor of London, Boris Johnson – that "some measure of inequality is essential for the spirit of envy and . . . is, like greed, a valuable spur to economic activity" (Johnson 2013). London, whose financial institutions played such a notable role in the financial crisis, does indeed show the pattern of inequality considered economically desirable by its mayor, with 16 percent of the citizens among the poorest tenth of the UK population nationwide and 17 percent in the richest tenth.[7] This trend has been tracked and reported on nationally by the Office of National Statistics:

> Between 1977 and 1991 the share of total disposable income received by the top fifth of households (or quintile group) increased from 36 to 42 per cent. The shares received by each of the lower three quintile groups fell, in the case of the bottom quintile group from 10 to 7 per cent.
>
> (Jones, Annan, and Shah 2008, 1)

We will return to the critical issue of inequality, but, for now, it is necessary only to observe an association between the rise of a certain idea of city competitiveness superficially expressed in creative and cultural terms and the dominant economic ideology of the past thirty years. A belief in unregulated competiveness – with increasing inequality as an unavoidable, and more or less undesirable, by-product – has become so normal as to pass almost unnoticed today. But it is not so obvious why the creative industries, and particularly cultural infrastructure and services, should have come to be seen as assets in the competitive advantage sought by urban leaders, planners, and theorists. The explanation is connected, at least in part, to the changing role of culture in the consumer societies that developed after the end of the Second World War and, with renewed vigour, since the 1980s. The most striking aspect of this is the huge growth in culture's importance during these decades, first in the developed world but gradually – and partly because of the globalization enabled by new information and communication technology – everywhere else.

At this point, it may be necessary to clarify what is meant by *culture* and *art*. Rather than the familiar distinction between artistic and anthropological concepts, I take culture to encompass everything human beings do to express their beliefs and values. Some of that is deliberate, but much of it is more or less unconscious: it is just the ideological stream in which all human beings swim, which shapes their thought and action and to which they contribute in turn – if they are able. And like a stream, it is fluid and changeable, so that what is acceptable today may be less so in ten years' time, and vice versa. In this characterization, art is a part of culture, but it has a distinct value as a creative toolbox that enables people to tinker knowingly with that culture – to create new expressions that symbolically and materially affirm, question, reposition, or otherwise engage with that cultural environment that surrounds and supports them. If culture is the expression of values, then art enables the construction (and deconstruction) of those values. It is how humans make and share meaning – privately among family and friends and publicly with strangers, though, again, the degree of consciousness involved is complex and variable. Culture and art have always been inseparable from what we currently think of as being human, but they are expressed differently within and between different groups, places, and times. That is why, for example, European and Australian Aboriginal concepts of the authorship, ownership, and purpose of art are often at odds with each other (Gennochio 2008).

One reason for the recent growth in the importance of art and culture in Western Europe is the parallel decline in authority experienced by other systems for creating meaning and expressing value, notably politics and religion.[8] The steady retreat in Western European societies of these framing narratives over the past thirty to fifty years has left culture in general and art in particular as the principal means of defining and expressing values, at least for now (Eagleton 2003). While religion retains a powerful influence in the United States and has grown in importance in other parts of the world, politics in the 1990s – after the prematurely announced "end of history" – has seemed often to be reduced to competing management styles. In Europe, the space vacated by religious and political belief has been occupied by culture and the cult of celebrity. At one level, this is evident in the massive growth of commercial arts production and consumption, itself facilitated by the globalized economy and its new technologies. Television, radio, film, music, books and magazines, fashion, digital and Internet-based media, computer games – the list of the creative class's preoccupations is long and familiar. The amount of time we spend consuming culture is unprecedented, enabled also by growing, if unequal, prosperity, education, and leisure in the West and across the world.[9] Veblen's observations about conspicuous consumption become more pertinent as the leisure class grows. The brand (i.e. commercial) value that culture can add to the consumer economy, from food and clothing to leisure, has been observed by various commentators, sometimes critically, sometimes not. Much of the competitive edge cities are now supposed to find is characterized not by *what* they make or do but by *how*.

It is not just commercial cultural production that has expanded during this period. European states have steadily increased their public investment in and support for arts and culture over the past thirty years. In Britain, a country with a historically sceptical view of the arts, the change has been striking. Between 1976 and 2006, the Arts Council's grant in aid from the national government increased by more than 400 percent in real terms – and this despite fears that the creation of a national lottery in 1994 would lead to a reduction of Treasury grant-in-aid.[10] In fact, in its first ten years the lottery brought an additional £2 billion to the arts (ACE 2005) – enough to refloat a small bank these days. Across the UK, cities have been endowed with new theatres, museums, concert halls, and public art. Redundant industrial buildings have been transformed into art galleries: Tate Modern receives about thirty thousand visitors a day on weekends and is already planning its second major enlargement since it opened in 2000. In Gateshead, in North East England, a new concert hall stands beside Baltic, a contemporary art gallery in a former flour mill. Similar regeneration projects can be seen in many deindustrializing European cities hoping to renew their economic and social fabric – in some cases, such as Liverpool, Lille, and Rotterdam, energized by the European Capital of Culture label. There is, however, a wide variation in the degree to which cultural action is integral to a sustained social and economic regeneration process rather than a symbolic gesture intended to civilize an external investment shaped by narrower commercial objectives.

Jamie Peck (2005) suggests that affordability and feasibility have made cultural investment particularly attractive to city leaders facing complex socioeconomic challenges in regional cities, with limited resources and control over their own decision making: "Investments in the 'soft infrastructure' of the arts and culture are easy to make, and need not be especially costly, so the creativity script easily translates into certain forms of municipal action" (749).

The idea that culture and creativity – the concepts are not always clearly differentiated – can drive urban development has itself become global. The Capital of Culture idea has been translated to the Arab world (in 2013 the title was awarded to Baghdad, and in 2014 to Tripoli), while Lima is Capital of Ibero-American Culture for 2014. The British Council has put the creative economy at the centre of its work for more than a decade, launching Creative City programmes for Eastern Europe and Southeast Asia in 2008 (see http://creativeeconomy.britishcouncil.org;

and http://creativecities.britishcouncil.org). Everyone, it seems, believes in the power of culture to revive the city.

This growth in cultural investment and activity will be welcome to those for whom culture is central to human society: my past research has often considered the value of participation in arts and culture in personal and collective development (Matarasso 1997). There can be little objection to investing in art, culture, and other creative activity as one aspect of modern urban policy, especially if it supplements rather than supplants wider established and novel approaches to social and economic development. But the rhetoric of creativity and openness should not prevent us from seeing the very real limits to who may be welcome in the cool contemporary cities promoted as Europe's urban future. To what extent, for example, do those five young asylum seekers, with their talents and their needs, have a place in the creative cities held up among the examples of liberal democracy that may have drawn them to the Netherlands in the first place?

This is not just an ethical or political problem, though it is both. It runs to the heart of whether these creative cities will be successful in the long term. The question is whether they can build sustainable futures by concentrating so much on the 30 percent of the population Richard Florida sees as belonging to the "creative class" rather than on all the inhabitants who, for Aristotle, make a city by and because of their diversity. In a 2009 post to his website, Florida writes that "it's important to recognize that states with large concentrations of working class jobs do very poorly in terms of wealth and well-being" (Florida 2009). But Richard Wilkinson and Kate Pickett propose another explanation for this correlation in their research into the social determinants of health. They use World Bank, United Nations, and OECD data to compare the official measures of health and social problems between twenty-one of the world's richest countries (Wilkinson and Pickett 2009).[11] They then cross-reference these indicators with data on income inequality within each country, using either UN data or the Gini coefficient. To test for the influence of cultural or other differences between societies, they use the same methodology to look at the fifty states in the United States. The result, whether the comparison is international or within the United States, is strikingly consistent. According to Wilkinson and Pickett, as far as prosperous societies are concerned, the degree of income inequality is directly linked to the incidence of a range of serious health and social problems. But they go further, arguing:

> These differences are not differences between high- and low-risk groups within populations that might apply only to a small proportion of the population, or just to the poor. Rather, they are differences between the prevalence of different problems which apply to whole populations.
>
> (173)

In other words, once a society has reached the level of prosperity now evident in the West, variations in life expectancy, ill health, crime, educational attainment, homicide, and other key indicators are accounted for, according to Wilkinson and Pickett, by the degree of internal inequality of each society. Countries such as Japan, Norway, or Sweden, where the income of the richest 20 percent is less than four times that of the poorest 20 percent, do consistently better in these critical life outcomes than countries such as the UK, Portugal, and the United States, where the differential is up to twice as great.

Wilkinson and Pickett are clear that the damaging effects of income inequality they identify operate primarily at a national rather than neighbourhood level. Yet people experience the negative aspects of inequality personally, where they live: the widening gap between the richest and the poorest is evident on British streets. As run-down Victorian city centres were transformed into smart spaces for retail, leisure, and culture in the boom years between 1995 and 2008, the role of private and commercial interests in the process began to be questioned. The forty-two-acre Liverpool One shopping and leisure centre, opened during that city's year as European Capital of Culture in 2008, has been criticised for a perceived privatization of public space. According to Roy Coleman, a sociologist at Liverpool University, "The rules for the newly privatised city centre fabricate an ideal citizen – aspirational in consumption and thinking big with urban pride" (quoted in Carter 2008). The Royal Institute of Chartered Surveyors has highlighted the risks associated with urban policies that passively or actively produce more segregated communities, observing that "while social exclusion is something 'they' have, polarisation is something we all suffer from" (Minton 2002, 32).

In Nottingham city centre, as in most British cities experiencing this kind of regeneration, it has become more and more difficult to find a cheap cup of tea with each passing year. There are plenty of global-brand coffee shops, but if you are young and unemployed, a branded coffee and the bus ride to buy it would consume a tenth of your week's welfare payment.[12] The reduction of space open to people without much cash in their pockets, with the cost of public transport, mean that the poor are increasingly absent from the glossy centres of many cities, except in destitution or homelessness. A sense of exclusion from the consumer lifestyle taken for granted by others was a factor in the riots that occurred in London, Manchester, Birmingham, Bristol, and elsewhere in August 2011 (Matarasso 2013). Making mid-scale cities attractive for the creative class – and especially those with money, education, and leisure – might strengthen their future competitiveness and prosperity, but if it is at the cost of increasing the disparity between the richest and the poorest and further embedding structural exclusion, the long-term benefits will be limited. If Wilkinson and Pickett are right about the effects of inequality, any benefit might be seriously compromised in any case. Some advocates of the present economic system are beginning to recognise this reality; in 2012, the *Economist* stated, in a widely discussed leading article, that "inequality has reached a stage where it can be inefficient and bad for growth" ("True Progressivism" 2012).

Such challenges require national and international policy responses. Perhaps some will emerge as a result of the present crisis, although there is little sign yet of imaginative thinking within the European political class, which seems unwilling to consider the implications of a deeper failure in recent economic policy (Stiglitz

2012). Be that as it may, individual cities still have scope to develop more inclusive and so ultimately more sustainable creative cultures. One positive step would be to recognise the risks of exclusion associated with existing cultural and urban strategies and seek ways to engage the population as a whole by making space for the full spectrum of their diverse interests.

Some years ago, the British Department for Culture, Media and Sport published a report on the role of the arts and sport in combating social exclusion (DCMS 1999).[13] Drawing on established arts and community development experience and theory, the "PAT 10" report made some sensible proposals designed, in the context of wider policy changes and interventions, to support the role of culture in reducing social exclusion. Some were adopted, with variable commitment and imagination, in the context of a general expansion of government resources for culture after 1999. But one – in its way the most inventive of them all – was largely ignored. The PAT 10 report proposed: "Consideration of experimental arrangements whereby funds are made directly available to accountable community groups . . . to spend on arts-related activity" (DCMS 1999, 60).

No connection was made between this idea and the government's interest, some years later, in community empowerment. The 2008 White Paper *Communities in Control* aimed "to pass power into the hands of local communities so as to generate vibrant local democracy in every part of the country and give real control over local decisions and services to a wider pool of active citizens" (DCLG 2008, 12). It made little reference to culture but proposed that Arts Council England and other cultural agencies should have a "duty to involve" local people in key decisions. These ideas were abandoned when the government lost power in 2010, and neither the role of culture in reducing social exclusion nor cultural inclusion – in the sense of empowering people as cultural actors – is on the present policy agenda. However, the ill effects of excessive inequality identified by Stiglitz, Wilkinson and Pickett, and others, to which either or both of these might be a partial response, have not diminished in the intervening years.

A degree of cultural inclusion might be advanced by building on the thinking of both the PAT 10 report and *Communities in Control* to give marginalised groups direct control over resources with which to plan, commission, and promote their own cultural activities. Instead of, or at least as well as, funding artists to work with community groups (as currently occurs), public authorities could be giving the people who currently benefit least from the creative transformation of their cities the means to be cultural actors in their own right, and not just the consumers, often no more than tolerated, of a metropolitan and globalised elite culture.

At the beginning of this century, a King Baudouin Foundation initiative, the Living Heritage programme (2000–2005), showed the potential of transferring control of cultural resources to community groups. Operating in four countries of Southeastern Europe in areas with little if any professional cultural infrastructure, Living Heritage made the most of the opportunity for direct support. About 140 communities received funds (usually for the first time) to develop cultural projects around their own needs, interests, and expertise. Where they required artists, curators, or other professionals to assist them, they set the terms and they paid the fees. As a result, when they did not agree with the expert advice, they had the confidence to reject it. Alongside the money, the communities got support and training: it would be naïve to suggest that such groups can be established and become successful without assistance, but they held the money and decided how it would be spent. Over five years, the programme invested €2.2 million in grants, training, and other support: 93 percent of the projects met or exceeded their agreed objectives (Matarasso 2005).

It is often said, particularly by those with no interest in a wider distribution of cultural resources or a greater diversity of artistic expression, that such policies lead only to poor quality arts work, worthy at best, but lacking the excellence that, by implication, is evident in the rest of the arts world. It is true that community arts work is not always good, even in its own terms. But neither is repertory theatre, contemporary dance, opera, or pop music. If the quality of an artist's work could be guaranteed, there would be no job for critics, and fewer poor artists. But in addition to its value in reducing exclusion through strengthening community empowerment, such direct cultural investment also offers an alternative dimension to the artistic life of cities today. At its best, by which any arts practice is entitled to be judged, this work is creatively, artistically, and culturally enriching precisely because it comes from other, usually unheard, people with different values and experiences to share.

Matt Peacock is a musician and opera critic who was also working in a London night shelter for homeless people in the late 1990s. Peacock was incensed by a throwaway comment by Sir George Young MP, then a Conservative Housing Minister, who had notoriously described the homeless as the "people you step over when you leave the opera," on BBC Radio 4's *Today* programme (29 June 1991). Determined to challenge the values and assumptions behind this remark, Peacock began singing workshops at the Passage Shelter, working towards a performance at the Royal Opera House's Linbury Studio in 2000. The experiment was a triumph for the homeless performers: Peacock says, "[I]t was the first time they'd felt respected, applauded, congratulated; they were very proud of it" (personal communication). But the concert was also a success with audiences, as its creators intended. They were not interested in developing performances that were appreciated because they were by homeless people: they wanted to create art with homeless people that had value in the everyday cultural life of London.

In the years since that first concert, Streetwise Opera has grown to work with homeless people in nine centres in London, Newcastle, and Nottingham. They have performed Britten's *Canticles*, Mahler's *Rückert Lieder*, and world premières of work by Will Todd and Ben Dunwell, Christina Jones and Orlando Gough, and Emma Bernard. They have sung in some of the UK's leading concert halls and received glowing reviews in the *Times*, the *Guardian*, and *Opera Now*. Above all, perhaps, they have created a distinctive aesthetic in pursuing their goal of ensuring that homeless people literally have a voice. They have ensured their inclusion in the city not through a condescending opera house outreach initiative but by creating formidable artistic work that empowers the performers culturally as well as socially. Streetwise Opera has shown that homeless people have something distinctive and valuable to contribute to the life of a creative city. Policies that prioritise an inclusive cultural life will not overcome the problems of inequality, but they may mitigate some of its effects and enable challenges to dominant cultures. They may even open space within which to rethink some existing assumptions about cities, societies, and cultures.

Streetwise Opera is artistically outstanding, but there are many other examples of inclusive creative practice.[14] There are similar projects throughout the UK, in France, in the Netherlands, and – as we have seen – in other European countries and, of course, beyond Europe. The problem is not a shortage of examples: the problem is their continuing marginalisation in contemporary cultural and urban policy. These projects go from hand to mouth, more often supported by social and charitable funds with remedial objectives than by arts funds that recognise the legitimacy of these distinctive voices. The consequence, whether intended or not, is to build a creative ghetto that parallels the educational, housing, health, and employment ghettos already provided for the poorest. Most cities rarely open their cultural flagships or their cultural discourse to these other voices with their distinctive values: still less do they offer the empowerment of resources. The self-representation of the poor and marginalised is tolerable, provided it remains on the far side of the ring road.

In *The Faraway Nearby*, Rebecca Solnit describes human experience as an ancient, limitless, and continuing conversation in which we are all made and to which we should all be able to add:

> You build yourself out of the materials at hand and those you seek out and choose, you build your beliefs, your alliances, your affections, your home, though some of us have far more latitude than others in all those things. You digest an idea or an ethic as though it was bread, and like bread it becomes part of you. Out of all this comes your contribution to the making of the world, your sentences in the ongoing interchange. The tragedy of the imprisoned, the unemployed, the disenfranchised, and the marginalised is to be silenced in this great ongoing conversation.
>
> (Solnit 2013, 191)

Dox Theatre, Streetwise Opera, and so many others show that those who are imprisoned, unemployed, disenfranchised, and marginalised – that is, those who have not done well in the competition that defines the contemporary European urban landscape – can still make a vital contribution to the making of the world. In the end, though, the inclusion of *all* voices within the city's agora is essential not just to a rich cultural life but also to the social and economic well-being of the whole community.

Notes

1 This chapter is developed from a keynote presented at Griffith University's Creative Communities Conference, in Surfers Paradise, Australia, on 16 April 2009.

2 *Het is Maar . . .* (*It's only . . .*) was developed for Dox by Jeanine Coco and Carole van Ditzhuyzen, with financing from Stichting De Vrolijkheid; see www.theatergroepdox.nl. I am grateful to Eugene van Erven, Senior Lecturer in Theatre Studies, Utrecht University, for information about the company and translations.

3 Richard Florida's ideas do not merit attention in Joseph Stiglitz's critique of recent American economic policy (Stiglitz 2012).

4 The United Nations reported that there were 145 countries with *fewer* than ten million inhabitants in 2013 (United Nations 2013); in 2011, the latest year for which there are published figures, there were twenty-three urban agglomerations with *more* than ten million inhabitants, a figure that is expected to have risen to thirty-seven by 2025 (United Nations 2012).

5 Globalization has also brought forms of bonded labour back to rich cities in the West and elsewhere, as millions of the world's poor expend their lives on the dirty jobs, remitting their pay, such as it is, to families in the developing or post-communist worlds. In late 2013, Amnesty International published a report into the human rights of migrant workers employed in facilities construction for the 2022 World Cup in Qatar, highlighting the relationship between global cultural events and exploitation (Amnesty International 2013).

6 In 2007, at what may come to be seen as the high-water mark of this period of art history, Damien Hirst sold a diamond-encrusted skull for £50 million. It remains to be seen whether the consortium of purchasers – rumoured to include the artist himself – will have struck a good deal. It is entirely characteristic of the current relationship between art and capital that the artist's latest speculations concern a 187-acre property development on the edge of Ilfracombe, the Devon town close to his country home (Philipson 2013).

7 Source: www.londonspovertyprofile.org.uk/indicators/topics/inequality/income-inequality-by-english-region/ [Accessed 13 January 2014].

8 The experience of Eastern European countries is different for obvious reasons, both before and after the end of Soviet Communism between 1989 and 1991, by which time the neoliberal economic policies adopted in the United States, the UK, Australia, Canada, and elsewhere were becoming orthodox.

9 The growth can be illustrated with reference to an area of cultural consumption often said to be in decline: cinema. In 1975, the UK had 1,547 screens that drew a total annual cinema audience of 118 million (McCosker 1999). In 2012, and despite the emergence of multichannel television and Internet-based entertainment, there were 3,817 programmed cinema screens being watched by an annual audience of 172.5 million people (BFI 2013). UK cinema audiences in 2013 were much lower than they were in the 1940s, when annual admissions peaked at 1.46 billion, but the British Film Institute estimated total audiences for film in 2012 at 5.1 billion on all platforms (television, DVD and Blu-ray, online, and cinemas); in 2013, the organisation reported that the diversity of formats made it impossible to reach a reliable estimate (BFI 2013).

10 The 2008 financial crisis has changed this situation, with Arts Council England's grant in aid being reduced by almost 40 percent between 2010 and 2014, while there have been similar sharp cuts in the Netherlands, Greece, Ireland, Italy, Hungary, and elsewhere, in addition to the effects of inflation.

11 They exclude states with fewer than three million inhabitants to avoid the distorting effects of tax havens.

12 Jobseeker's Allowance (JSA) is the main UK benefit for people of working age who are out of work. A single person is currently entitled to £56.80 per week (aged eighteen to twenty-four) or £71.70 (over twenty-five years old); some people may also be eligible for welfare assistance with rent and some other costs.

13 In 1998, I was member of the committee (unpaid) that contributed to the report, my first and last involvement in government policy making.

14 Even in the relatively narrow field of music and homelessness, there are several well-known projects, including the Accueil Bonneau Choir (Montréal, Canada, now disbanded), the Sydney Street Choir (Sydney, Australia, formed in 2001), and the Choir of Hard Knocks (Melbourne, Australia, formed in 2007).

References

Amnesty International (2013) The Dark Side of Migration: Spotlight on Qatar's Construction Sector Ahead of the World Cup. London: Amnesty International. Available at: www.amnesty.org/en/news/qatar-end-corporate-exploitation-migrant-construction-workers-2013-11-17 [Accessed 6 January 2014].

Aristotle (1992) *The Politics*. Translated by T. A. Sinclair and revised by T. J. Saunders. London: Penguin.

Arts Council England (ACE) (2005) *Transforming the Cultural Landscape: Why the Lottery Is Good for the Arts*. London: ACE.

British Film Institute (BFI) (2013) *Statistical Yearbook 2013*. Available at: www.bfi.org.uk/education-research/film-industry-statistics-research/statistical-yearbook [Accessed 6 January 2014].

Carter, H. (2008, May 28) Policing the retail republic. *The Guardian*. Available at: www.guardian.co.uk/society/2008/may/28/regeneration.communities [Accessed 12 January 2014].

Chang, H.-J. (2010) *23 Things They Don't Tell You about Capitalism*. London: Penguin.

Department of Communities and Local Government (DCLG) (2008) *Communities in Control: Real People, Real Power*. London: DCLG.

Department of Culture, Media and Sport (DCMS) (1999) *Policy Action Team 10: A Report to the Social Exclusion Unit*. London: DCMS.

Eagleton, T. (2003) *After Theory*. London: Penguin.

Florida, R. (2002) *The Rise of the Creative Class and How It's Transforming Work, Leisure and Everyday Life*. New York: Basic Books.

Florida, R. (2008) *Who's Your City? How the Creative Economy Is Making Where to Live the Most Important Decision of Your Life*. New York: Basic Books.

Florida, R. (2009, March 17) Class and well-being [blog post]. Available at: www.creativeclass.com/creative_class/2009/03/17/class-and-well-being/ [Accessed 12 January 2014].

Gennochio, B. (2008) *Dollar Dreaming: Inside the Aboriginal Art World*. Prahan: Hardie Grant.

Johnson, B. (2013) The 2013 Margaret Thatcher Memorial Lecture, Centre for Policy Studies, 27 November. Available at: www.cps.org.uk/events/q/date/2013/11/27/the-2013-margaret-thatcher-lecture-boris-johnson/ [Accessed 13 January 2014].

Jones, F., Annan, D., and Shah, S. (2008) The distribution of household income 1977 to 2006/07. *Economic and Labour Market Review*, 2(12). London: Office for National Statistics.

Landry, C. (2000) *The Creative City: A Toolkit for Urban Innovators*. London: Earthscan.

Landry, C., and Bianchini, F. (1995) *The Creative City*. London: Demos.

Matarasso, F. (1997) *Use or Ornament? The Social Impact of Participation in the Arts*. Stroud: Comedia.

Matarasso, F. (2005) *Living Heritage: Community Development Through Culture Resources*. Brussels: King Baudouin Foundation.

Matarasso, F. (2013) 'All in this together': The depoliticisation of community art in Britain, 1970–2011. In E. van Erven (ed.), *Community, Art, Power* (pp. 214–240). Rotterdam: International Community Arts Festival (ICAF).

McCosker, P. (1999) A brief history of cinema exhibition in the UK. Available at: www.mediasalles.it/crl_cosker.htm [Accessed 6 January 2014].

Minton, A. (2002) *Building Balanced Communities: The US and UK Compared*. London: Royal Institution of Chartered Surveyors (RICS).

Peck, J. (2005) Struggling with the creative class. *International Journal of Urban and Regional Research*, 29(4), 740–770.

Philipson, A. (2013) Damien Hirst submits plans to build entire town in Devon. *The Telegraph*, December 8. Available at: www.telegraph.co.uk/news/celebritynews/10504296/Damien-Hirst-submits-plans-to-build-entire-town-in-Devon.html [Accessed 6 January 2014].

Sandel, M. (2012) *What Money Can't Buy: The Moral Limits of Markets*. London: Penguin.

Solnit, R. (2013) *The Faraway Nearby*. London: Granta.

Stiglitz, J. (2012) *The Price of Inequality*. London: Penguin.

True progressivism (2012, October 13) *The Economist*. Available at: www.economist.com/node/21564556 [Accessed 12 January 2014].

United Nations, Department of Economic and Social Affairs, Population Division (2012) *World Urbanization Prospects: The 2011 Revision*. ST/ESA/SER.A/322.

United Nations, Department of Economic and Social Affairs, Population Division (2013) *World Population Prospects: The 2012 Revision, Highlights and Advance Tables*. Working Paper No. ESA/P/WP.228.

Wallerstein, I. (2008) The demise of neoliberal globalization. *Yale Global Online*. Available at: http://yaleglobal.yale.edu/content/2008-demise-neoliberal-globalization [Accessed 8 January 2014].

Wilkinson, R., and Pickett, K. (2009) *The Spirit Level: Why More Equal Societies Almost Always Do Better*. London: Allen Lane.

Wood, P., Landry, C., and Bloomfield, J. (2006) *Cultural Diversity in Britain: A Toolkit for Cross-Cultural Co-operation*. York: Joseph Rowntree Foundation.

10 Cultural access and activation

Civic participation in local sustainable communities

Anita Kangas and Sakarias Sokka

Introduction

The concept of sustainable development contains economic, social, cultural, and ecological dimensions. According to Throsby, within the framework of sustainability, culture can be viewed in both a functional and a constituent sense. The former includes arts, cultural products, and cultural industries, whereas the latter includes group values, customs, traditions, beliefs, values, and identity (Throsby 2001). As there are systematic differences in how people do, make, and say things, there also are different cultures (cf. Cahoone 2005). Lily Kong (2009, 3) has defined *cultural sustainability* as "the ability to create local cultural content, embed indigenous idioms in cultural products, and, possibly, devise unique cultural forms that underscore a local sense of identity and indeed, nationhood, particularly in the face of globalizing and potentially homogenizing forces." The concept of cultural sustainability highlights the ability of culture to "forge a productive diversity for the human species" as well as to "nurture the sources of cohesion and commonality," recognising culture to be "the glue of similarity (identity; literally) that grounds our sociability" (Kong 2009, 3).

Cultural and social sustainability can be seen as closely intertwined. According to Kong, *social sustainability* means healthy social interaction, protection of the vulnerable, and respect for social diversity. It calls for systems, structures, and programmes that allow "our participation as autonomous yet social beings" (Kong 2009, 3) and emphasises the social dimension of cultural activities whereby a socially sustainable cultural policy/activity is one that enables social inclusion and the building of community bonds. It is important to sustain the cultural traditions, beliefs, values, and fundamental convictions that constitute individual and collective identity within the limits of universal human rights. In the same way, preserving social institutions that govern production, education, political stability, and social coherence is essential.

An external norm arising out of thinking about sustainability is its ethical motivation – sustaining quality of life means preserving the elements of natural, social, and cultural capital that humans appreciate today as options for future generations. Thus, its justification is the principle of intergenerational equity (Throsby 2001). In terms of social equity, accessibility is a fundamental measure.

It follows that the built environment and transport – for example, the key services and facilities – can have an impact on the extent and nature of accessibility in a given place. The aspects of everyday life to which residents and users need equitable access include education and training, decent housing, public services, (social) infrastructure, green space, culture, and recreation (Dempsey, Bramley, Power, and Brown 2011).

In this chapter, we are interested in access to, and participation in, culture as a part of quality of life and cultural sustainability. Participation is examined from an individual perspective as well as a structural one. An individual's activities are the starting point in participation, but equally in focus are the frames that enable this participation, the societal mechanisms related to power and the exercise of power. Based on these premises, we review two different projects that make it possible for us to ask: (1) Who has access to cultural activities in Finnish municipalities, and who/which groups are excluded? (2) How can cultural administrators and artists enable access, encourage inhabitants to participate in cultural life, and generate sustainable welfare in local communities?

The case of Finland

Cultural policy became institutionalised in Finland during the 1950s, at the same time as in other Nordic countries. During the 1970s and 1980s, the cultural sector was established as one of the public sector services. The division of responsibilities between the state and local government was built up like other welfare state services: local government has a crucial role in delivering and providing services, and the state regulates them through legislation and financial support. Cultural sectors cooperate with many other municipal sectors and voluntary associations. At the end of the 1980s, the ideas of New Public Management took root in Finland, and soon after came the financial recession of the 1990s (Honkapohja and Koskela 1999). Together, they changed the position of the cultural sector, especially in small municipalities (under twenty thousand inhabitants) (Kangas 2003, 2004; Kangas and Vestheim 2011).

According to Kangas and Kivistö (2011), various forms of municipal diversification constitute a major challenge for local cultural activities because they have made the availability of cultural services more unequal. Public actors face conflicting pressures: services should be implemented efficiently and more cheaply but at the same time increasingly take into consideration consumers' freedom of choice and needs. Furthermore, the inequalities in people's leisure time have become more evident during recent decades, when the role of the citizen has predominantly turned into that of consumer and buyer. In this difficult context, community-based and participatory approaches have become associated with the role of art, creativity, and cultural activities for community vitality and community planning.

Our analyses are based on original research findings from two sequential research projects conducted at the Unit of Cultural Policy of the University of Jyväskylä since 2011. The first project mapped people's arts- and culture-related free time activities, and it provided much-needed information about problems in accessing and participating in cultural services. The second project is a participatory action research project, still in process, in which research is harnessed to find solutions to problems in accessing and participating in culture.

Data and methodologies

The first research project, *Local Cultural and Sport Services: Access and Democracy* (CULTSPO), was based on interviews and survey results, both conducted in 2012. We analysed the interviews ($N = 167$) that were conducted in a small Finnish city (twenty thousand inhabitants) and the responses to an Internet-based survey ($N = 675$). For the survey, we collected replies from citizens who had experience in using local cultural services and were therefore capable of identifying problems in access and participation. The interviews provided more precise knowledge about people's motivation in using cultural services. We analysed the survey results with logistic regression analysis (see Table 10.1) and the interviews using content analysis (see Table 10.2).

The second project (launched by the Ministry of Education and Culture), *Activating Cultural Participation in Local Communities* (KUULTO), is an action research project that asks how to generate new and sustainable practices for the production of cultural services and how to encourage people to participate. It focuses on experiments being conducted in different parts of Finland between 2012 and 2015 by municipal cultural departments, associations, and a private company. The project deliberately focuses on municipalities in which the net costs for general cultural activities per resident are very low. The first call for funding of projects resulted in 188 applications, of which twenty-two were selected according to considerations such as the novelty value of the idea, the potential for increased access to cultural activities, the versatility of the activities, the existence of strong cooperation, the idea's feasibility, and the sustainability of the operational models. As always in action research, the project also aims to change social reality although it is never certain that change will actually take place, and the change may differ from the original goal.

Participation in local cultural activities is unequal

The promotion of equality is a central aim in Finland's cultural policy. According to the CULTSPO survey, however, there are big differences in different groups' ability to access cultural services in Finland.[1] As shown in Table 10.1, social stratification cannot be overlooked in the use of cultural services. If you are a woman and have an academic degree, are older than thirty-five (and, especially, older than fifty-five) years of age, live in a city, and earn over €40,000 annually, you are more likely to make four or more visits per year to *at least four* of the following: art galleries, cinema, dance performances, art/culture festivals, museums, libraries, opera, orchestras, and/or theatres. Women also have more culture- and art-related hobbies than men.

Table 10.1 Access to cultural activities and feedback about cultural services (Internet survey)

	Active visitor to art gallery, cinema, dance performance, art/culture festival, museum, library, opera, orchestra, and/or theatre	*Regular and active art/culture related hobbies*	*Has given feedback on local cultural services*
Sex			
Woman	1.00	1.00	1.00
Man	**0.36*****	**0.41*****	0.86
Age			
18–34	1.00	1.00	1.00
35–54	**1.18***	0.97	**1.82***
Over 55	**2.06****	0.67	**5.69*****
University education			
No	1.00	1.00	1.00
Yes	**2.76*****	**1.70****	1.27
Lives in			
City	1.00	1.00	1.00
Rural area	0.70	1.14	**1.46***
Children live in the same household			
Yes	1.00	1.00	1.00
No	1.40	**0.63***	0.94
Income per year			
< €20,000	1.00	1.00	1.00
€20,001–30,000	0.99	**0.50****	0.75
€30,001–40,000	1.03	**0.45****	0.92
> €40,001	1.65	**0.42****	0.83
Nagelkerke R^2	0.16	0.12	0.13
N = 656			

*** $p < .001$

** $p < .01$

* $p < .05$

Further, regardless of gender, not having children increases the likelihood of making visits. This relates both to the flexibility of time use that people without children have and to the higher cost of making visits with children. On the other hand, having children and/or low annual income increases the likelihood of having hobbies: children are more likely to have culture-related hobbies than adults, and hobbies are often cheaper than making regular visits to cultural institutions.

People over fifty-five do not have as many active hobbies as younger cohorts, but they are more active in making visits. Thinking about the organisation of cultural services, it is especially interesting that the older cohorts are much more eager to give feedback about their local services than younger people. Also, people living in rural areas are relatively active in giving feedback, even though they usually have far fewer opportunities to attend cultural events and visit art facilities in their own localities than city dwellers. It is perhaps easier to give feedback in smaller communities where the services are closer to inhabitants; it might also be that lack of services produces feedback.

Our results match up well with the results of earlier studies, according to which middle-aged, well-educated women are the most active visitors; higher education and higher income predict a higher number of visits; middle-aged people are more likely to visit the theatre and attend concerts; and having no children in the household increases the likelihood of making visits to art and culture facilities and performances (e.g. Chan and Goldthorpe 2005; Christin 2012; DiMaggio and Mukhtar 2004; Galloway 2006; Liikkanen 2005; Minkkinen, Pääkkönen, and Liikkanen 2001; Purhonen, Gronow, and Rahkonen 2011; Stanbridge 2011; Suominen 2013; Van Eijck and Knulst 2005).

Compared to the rest of the EU, Finland and Scandinavia in general have the lowest average number of people who have not used any cultural services in twelve months. According to an EU-wide comparison of thirty-four countries, the number of residents who have used cultural services more than twelve times in the same period is highest in Finland (Eurostat – European Commission 2011). Even so, residents' opportunities to consume these services are not equal across Finland. Resources for cultural activities are unevenly distributed in Finnish municipalities because their willingness to invest in this area varies significantly (Kangas and Ruokolainen 2012). This observation is associated with the level of service provision and inhabitants' opportunities to participate in the activities. According to the data analysed by Statistics Finland, regional differences between Finland's densely and sparsely populated areas in the consumption of culture are among the highest in the EU.

Defining barriers to access

Our CULTSPO interviewees in a small Finnish city gave four sets of reasons for visiting culture facilities and/or having art- and culture-related hobbies: one reason was to improve their own competences,[2] another was connected to individual lifestyle, a third was to cement social identity, and the fourth – by far the most popular reason – was to improve mental and psychic well-being. The position of well-being was the same regardless of whether people were active in making visits to cultural facilities and art events, or in pursuing art- and culture-related hobbies, or both. Considering the individual's level of well-being, the particular mode of culture-related activities is not as important as simply being active in some mode or another. Furthermore, the positive effects of cultural services reach beyond the individual level. In our Internet survey, no less than 97 percent of respondents saw cultural services as important in increasing well-being and reducing social problems more broadly (Sokka et al. 2014).

Half (50 percent) of the interviewees in a small Finnish city were willing to increase their activity in hobbies related to culture and sport. They also offered many reasons for why they were unable to do so, and these were almost the same reasons our respondents to the e-survey gave for not participating in local leisure services. The biggest obstacles to access were (1) content of the services is too established or "not directed to me" (35 percent of respondents mentioned this), (2) opening hours are too restricted (30 percent), (3) prices are too high (26 percent), and (4) services are poorly situated (24 percent) (Sokka et al. 2014, 57).

Local residents also felt they were unable to influence municipal decision making about the development of cultural activities. In our e-survey, a mere 23 percent of the respondents said that people in Finland have good opportunities to influence service provision (Sokka et al. 2014).[3] People do not see themselves as having much power to remove barriers to accessing services (see Table 10.2).

According to respondents, municipal officers have the most power to influence culture services. Kangas and Ruokolainen (2012) have reported that the staff of municipal cultural departments recognise inequalities in access to cultural activities: the elderly (in institutional care and far from city centres), young people, individuals with a physical or other disability, people living in remote villages and suburbs, people with a low income, the unemployed, people with a low educational level, and immigrants all have difficulties using cultural services. The barriers that the administrators identified were thus related to region, social status, language, age, and use of time. A third of them would like to bring cultural services to remote villages, suburbs, and social and health care institutions. They would also like to see a more positive attitude in cities toward local grassroots projects, but lack the means to improve the current situation.

Table 10.2 Attitudes on power in cultural and sport services (Internet survey)

Who in your municipality has the most power to influence culture and sport services? (mention the three most important in order from 1st to 3rd) (N = 675)	*Mean (1–3)*	*No. of times mentioned among the three most important*	*Standard deviation*
1. Municipal officers	1.58	517	.742
2. Local sporting clubs and sport associations	1.87	425	.773
3. Members of the municipal executive board	1.89	263	.768
4. Members of the municipal council	1.91	288	.782
5. Members of the sport and/or cultural committee	1.92	301	.726
6. Mayor	1.97	204	.815
7. Cultural and art associations and societies	2.06	269	.741
8. Private companies	2.28	221	.747
9. Citizens	2.40	208	.817
10. Residents' associations	2.41	167	.623

Encouraging inhabitants

The CULTSPO project showed that municipal structures are not organised in the best possible way to enable social inclusion, build community bonds, or enhance people's participation "as autonomous yet social beings" (Kong 2009, 3). When the current economic situation and administrative changes force municipalities to restructure their services, the restructuring seems to be carried out without reference to the feedback that people give (or, in many cases, would be willing to give but are not asked). As a consequence, the need for new kinds of participatory approaches is justified.

The aim of the KUULTO project was to increase participation in cultural activities. Different motivating measures were designed for those people who had not been active before. Increasing participation also implied encouraging these individuals to take part in making decisions about activities and evaluating them. The main idea was to lead them to be committed to the activities and their development. The project made use of the action research approach and the concept of *participation* based on the theories of Paulo Freire (1972) and Augusto Boal (2002). Freirean participation is defined as a dynamic, interactional, and transformative process of dialogue between people, groups, and institutions that enables people, both individually and collectively, to realize their full potential and be engaged in their own welfare. Principles of dialogue, interaction, problem posing, reflection, and conscientization are important.

From this perspective, participation requires a new way of thinking from the actors in the experiment – that is, from the staff of cultural departments, residents, and those who evaluate the experiment. The question is how knowledge and information that can be gathered via resident participation and the development of activities could be taken into consideration within the operation (planning, budgeting, and administration) of the various municipal departments (Kangas, Jakonen, and Havimäki 2014). Municipal officers are experts who possess the knowledge generally needed for service provision. When local residents participate, they add experience and value-based knowledge to the processes. Their inclusion in development work should be a given, particularly in strategic and operational areas. In addition to asking "what?" and "where?" one also needs to ask "what kinds?" and "in what way?" (Sipilä, Bäcklund, and Tyrväinen 2009, 48; see also Grodach and Silver 2012; Parker 2011).

As action research, the KUULTO project asked the core question, how have the needs of local residents been heard and interpreted? According to the responses received, the "customer-oriented way of working," which is emphasized in all internal training in municipal administration, had been adopted at least rhetorically (see Clarke et al. 2007) and was highlighted by all the participants in the experiments. Service providers emphasised that if services are provided exclusively based on a top-to-bottom principle, local residents do not necessarily adopt the activities as integral parts of their lives.

At the organisational level, a traditional top-to-bottom bureaucracy presents obstacles to empowerment-based participation. Braye and Preston-Shoot (1995) listed some barriers they experienced in empowerment processes.[4] One barrier relates to the organisation of services and the administration's relationship with local residents. The experts fear that they may lose their status and power, that their professional skills might not be adequate, and that their expertise may be denied. They are suspicious about the clients' readiness and emerging competence and emphasise the mechanisms that reinforce (jurisdictional) power through legislation and administrative terminologies. A common claim is that there is no money for the necessary changes. On the other hand, local residents may question the experts' motivation and their authenticity. Residents are afraid of change and insecure about what to expect after a potential change. Therefore, it is difficult for them to join the processes.

Table 10.3 describes how the KUULTO project aimed to increase the accessibility of culture and break down barriers to participation. Barriers were organised into four categories: structural barriers, barriers to finding ways to have an influence, geographic barriers, and social barriers. The results showed how successful or difficult each process was in removing the barriers (Kangas et al. 2014). The project also identified key challenges related to the various efforts that are expected from different administrative and political actors at all levels.

In attempting to overcome the barriers and improve knowledge-sharing practices, the KUULTO project found that differing demands for information call for different data-collection methods (Kangas et al. 2014). For example, the use of "cultural probes" was a fruitful method of gathering inspirational data about people's lives, values, and thoughts. The probes are small packages that can include a variety of artefacts (e.g. maps, postcards, pictures, camera, and diary) along with evocative tasks, which are given to participants to allow them to record specific events, feelings, or interactions. The aim of using such probes was to obtain responses from families and thus stimulate the imagination of those responsible for organising cultural activities for them (see Crabtree et al. 2003; Gaver, Dunne, and Pacenti 1999; Gaver et al. 2004).

Another successful way to encourage people to participate was discovered in one KUULTO suburb. The idea was to motivate people by providing them with new opportunities to share their memories. "Culture" was not exported to suburban areas; instead, the already existing yet largely hidden culture and heritage was strengthened. Residents told stories related to the naming of places in that particular area and specific to the local people. Cooperation between generations injected new perspectives into the debate: for example, new designations given by children concerning places that were important to them gave them new meanings for the whole community. This kind of transmission and slow transformation of knowledge between generations is important for cultural sustainability.

A related idea was the "box for memories" that makes the creator's memories visible and that links memories across generations. The box was a collection of objects related to an individual's life, linked to time, travel, work, and so on. Discussions based on the boxes strengthened the connection between generations and between community members. Choosing various elements to put into the box of memories also enriched relationships among family members. These

Table 10.3 An overview of the first cycle of the KUULTO action research process and its main findings

Types of barriers	*Actions taken by the KUULTO project to overcome barriers*	*Challenges*
Structural barriers	• Collaboration with other municipal sectors (cultural and social, health, education, planning) • Collaboration with the third sector • Project funding; new resources and hiring artists • Action research project	• Problems in collaboration between municipal administrative sectors • Discontinuity of project funding • Problems in finding the right places for artistic work • After the action research project, how to continue reflexive bottom-up processes? How does the education system of professional artists respond to new kinds of demand? • What happens to newcomers?
Barriers to finding ways to have an influence	• Outreach cultural work • Social media: Internet and blogs • Working groups for people at the grassroots level (associations, villages) • New ways to empower citizens: civic forums and panels, cultural probe as a method, feedback and assessments	• Too much control and top-down practices • The directors are not able to delegate and do not allow younger people to make their voices heard • How to select members of working groups? • How can citizens' experience-based influence be transmitted to the planning and administrative processes of local government?
Geographic barriers	• Cultural activities and performances in villages and suburbs • Culture-bus and travelling performances • Collaborations with art institutions; bus transport to the performances • Professional artists mentoring, activating, and making interventions in villages and suburbs • Community artists; cultural companions	• How to secure continuity for artists' work in small and remote locations? • Are there enough artists who would be interested in community art? • How could art institutions become more interested in overcoming geographic barriers?
Social barriers	• Cultural activities in retirement homes, hospitals, and kindergartens • Cultural activities for elderly people who live in their own homes • Cultural activities in urban marginalised communities and in rural and outlying areas • Cultural activities to promote well-being in workplaces	• Impermanence in the cultural services • Getting directors of various sectors committed to new activities • Finding a common language between artists and other professionals • How to find new partners? • Tight schedules in social sectors' institutions and various workplaces

memory boxes could also be used by healthcare professionals with those for whom they care.

In another situation, community artists in KUULTO used Augusto Boal's (2002) "theatre of the oppressed" method, using theatre as a vehicle for participatory social change, and it proved to be very successful in removing social barriers. If we think about local cultural policy development, one of the most important questions that arose from this experience is how municipalities could employ artists and enable them to work with marginalised groups on a more permanent basis.

Conclusions

This chapter has investigated empirically the relation between participation in cultural services and barriers to accessing cultural activities. It involved a critical examination of the state of the art in local cultural policy and service provision in Finland. The overall purpose was to develop a conceptual foundation for investigating cultural participation. This involved applying the concepts of quality of life and cultural sustainability (including intergenerational equity, social equality, and empowerment) to the initial results of two research projects (CULTSPO and KUULTO) that are being conducted at the University of Jyväskylä.

According to these results, it is important to sustain the cultural traditions, beliefs, values, and fundamental convictions that constitute individual and collective identity within the limits of universal human rights and the demands of a global economy. These elements of culture are linked to the idea of sustainability in any local community that deals with questions like access, participation, and cohesion (see Dempsey et al. 2011). After all, without culture there would be no base for sharing the meanings that form communities. Cultural sustainability includes actions and issues that affect how communities manifest identity, preserve and cultivate traditions, and develop belief systems and commonly accepted values.

Yet much still remains to be done to foster cultural participation in municipal administration. There are large differences in different groups' access to and participation in cultural services. People recognise flaws in contemporary service structures that prevent participatory activities from developing but feel themselves powerless to influence the reorganisation of services. It is not impossible to remove the regional, social, and structural barriers that we have identified, but this cannot be done without new ways of organising the services.

Preserving social institutions that govern production, education, political stability, and social coherence is vital. However, lowering the walls of (cultural) institutions is just as crucial. For example, organising happenings outside the doors of institutions entails a new kind of professional collaboration between established artists and municipal officers. Strengthening participation also calls for new types of funding structures that could provide community artists with a reasonable salary. Furthermore, organising cultural activities with civil society actors in different settings would require more stable project funding and long-term strategies. Encouraging people to take part in developing their communities also calls for the acceptance of new actors in roles that have been occupied until now by those with identified administrative expertise in various areas.

Cultural facilities are fundamental for communities in promoting their culture, creativity, cohesion, and sustainability. Artistic processes play important roles in the societal transition to a more sustainable basis. Artists can create new spaces for dialogue and enable new ways of thinking, communicating, and building competencies that link cognitive and experiential insights. They catalyse shifts of societal consciousness, increasing both awareness and knowledge – collectively contributing to the transformation of individuals and communities over time.

Culture is a good in itself that needs to be preserved as the main component of personal and collective identity. Cultural sustainability as a normative concept relates art, creativity, and cultural activities to community vitality and community planning. Cultural beliefs and values are prime motivational agents for individual and collective actions; they co-determine the way that humans act in favour of or against sustainability. If the preservation of certain goods is not embedded in the portfolio of what a culture prescribes as valuable, any attempt to enforce this preservation will be futile in the long run, even if force is being used. As much as we need to preserve natural, economic, and social capital, we are also in urgent need of nurturing the cultural capital that provides the basic motivation and ontological security for human beings.

Notes

1 Our logistic regression analysis demonstrates which factors in the e-survey data predict, first, active visits to cultural facilities and art events (i.e. at least four visits during the last twelve months); second, engaging in active art- and culture-related hobbies at least once a month (i.e. respondent does at least four art and culture related hobbies: reads a book; plays a musical instrument; sings, draws, or paints; writes poems, novels, or short stories; takes photographs; and/or dances); and, third, giving feedback about local cultural services.
2 This however was not as much linked to success in working life (cf. Stebbins 2012) as in cultivating oneself and developing one's own skills for personal purposes.
3 Only 3 percent of respondents listed voting as among the best ways to affect service provision.
4 The original meaning of the term *empowerment* was political and related to activities against oppression (Freire 1972).

References

Boal, A. (2002) *Games for Actors and Non-actors*. New York: Routledge.
Braye, S., and Preston-Shoot, M. (1995) *Empowering Practice in Social Care*. Buckingham, PA: Open University Press.
Cahoone, L. E. (2005) *Cultural Revolutions: Reason Versus Culture in Philosophy, Politics, and Jihad*. University Park: Pennsylvania State University Press.
Chan, T. W., and Goldthorpe, J. H. (2005) The social stratification of theatre, dance and cinema attendance. *Cultural Trends*, 14, 193–212.

Christin, A. (2012) Gender and highbrow cultural participation in the United States. *Poetics*, 40, 423–443.

Clarke, J., Newman, J., Smith, N., Vidler, E., and Westmarland, L. (2007) *Creating Citizen-Consumers: Changing Publics and Changing Public Services*. London: Sage.

Crabtree, A., Hemmings, T., Rodden, T., Cheverst, K., Clarke, K., Dewsbury, G., and Rouncefield, M. (2003) *Designing with Care: Adapting Cultural Probes to Inform Design in Sensitive Settings*. Proc OZCHI. Available at: www.cs.nott.ac.uk/~axc/work/OzCHI03.pdf [Accessed 30 October 2014].

Dempsey, N., Bramley, G., Power, S., and Brown, C. (2011) The social dimension of sustainable development: Defining urban social sustainability. *Sustainable Development*, 19, 289–300.

DiMaggio, P., and Mukhtar, T. (2004) Arts participation as cultural capital in the United States, 1982–2002: Signs of decline? *Poetics*, 30, 169–194.

Eurostat – European Commission (2011) *Cultural Statistics*. Eurostat pocketbooks. Luxembourg: Office for Official Publications of the European Communities.

Freire, P. (1972) *Pedagogy of the Oppressed*. Harmondsworth: Penguin.

Galloway, S. (2006) Cultural participation and individual quality of life: A review of research findings. *Applied Research in Quality of Life*, 1, 323–342.

Gaver, W., Boucher, A., Pennington, S., and Walker, B. (2004) Cultural probes and the value of uncertainty. *Interactions*, 11, 53–56.

Gaver, W., Dunne, T., and Pacenti, E. (1999) Design: Cultural probes. *Interactions*, 6, 21–29.

Grodach, C., and Silver D. (2012) *The Politics of Urban Cultural Policy: Global Perspectives*. Abingdon, Oxon: Routledge.

Honkapohja, S., and Koskela, E. (1999) The economic crisis of the 1990s in Finland. *Economic Policy*, 29, 401–436.

Kangas, A. (2003) Culture as a local tool and public service. *Kunnallistieteellinen aikakauskirja*, 31, 332–340.

Kangas, A. (2004) New clothes for cultural policy. In P. Ahponen and A. Kangas (eds.), *Construction of Cultural Policy* (pp. 21–42). SoPhi 94. Jyväskylä: Minerva.

Kangas, A., Jakonen, O., and Havimäki, S-M. (2014) *Osallistumista ja aktivointia. KUULTO-toimintakokeilun matkassa*. Opetus- ja kulttuuriministeriön julkaisuja 2014: 11. Helsinki.

Kangas, A., and Kivistö, K. (2011) *Kuntien kulttuuritoiminnan tuki- ja kehittämispolitiikka*. Available at: www.minedu.fi/OPM/Julkaisut/2011/liitteet/tr12.pdf [Accessed in September 2013].

Kangas, A., and Ruokolainen, V. (2012) *Toimintamalli muutoksessa. Tutkimus kuntien kulttuuripalveluista*. Helsinki: Kulttuuripoliittisen tutkimuksen edistämissäätiö Cupore.

Kangas, A., and Vestheim, G. (2011) Institutionalism, cultural institutions and cultural policy in the Nordic countries. *Nordisk kulturpolitisk tidskrift*, 13, 132–151.

Kong, L. (2009) Making sustainable creative/cultural space in Shanghai and Singapore. *Geographical Review*, 99, 1–22.

Liikkanen, M. (2005) Yleisönä – kodin ulkopuolella ja kotona. In M. Liikkanen (ed.), *Suomalainen vapaa-aika. Arjen ilot ja valinnat* (pp. 67–100). Helsinki: Gaudeamus.

Minkkinen, V., Pääkkönen, H., and Liikkanen, M. (2001) *Kulttuuri- ja liikuntaharrastukset 1991 ja 1999*. Helsinki: Tilastokeskus.

Parker, S. (2011) *Cities, Politics and Power*. Routledge Critical Introductions to Urbanism and the City Series. New York: Routledge.

Purhonen, S., Gronow, J., and Rahkonen, K. (2011) Highbrow culture in Finland: Knowledge, taste and participation. *Acta Sociologica*, 54, 385–402.

Sipilä, M., Bäcklund, P., and Tyrväinen, L. (2009) Vaikuttamisen rajoilla: Onko koetulla luonnolla sijaa kaupungin suunnittelussa ja päätöksenteossa? *Alue ja ympäristö*, 38(1), 39–50.

Sokka, S., Kangas, A., Itkonen, H., Matilainen, P., and Räisänen, P. (2014) *Hyvinvointia myös kulttuuri- ja liikuntapalveluista*. Tutkimusjulkaisu 77. Helsinki: KAKS–Kunnallisalan kehittämissäätiö.

Stanbridge, A. (2011) The tradition of all the death generations: Music and cultural policy. *International Journal of Cultural Policy*, 13, 255–271.

Stebbins, R. A. (2012) *The Idea of Leisure: First Principles*. New Brunswick; London: Transaction Publishers.

Suominen, S. (2013) Essays on the demand for cultural performances. Acta Wasaensia 287, Taloustiede 8. Vaasa.

Throsby, D. (2001) *Economics and Culture*. Cambridge: Cambridge University Press.

Van Eijck, K., and Knulst, W. (2005) No more need for snobbism: Highbrow cultural participation in a taste democracy. *European Sociological Review*, 21, 513–528.

11 Provincial Poland

Sustainable development and culture in small and medium-sized towns

Katarzyna Plebańczyk

Introduction

Over the past twenty-five years, Poland has undergone profound changes due to the political transformation after 1989 and the EU accession of 2004. Sustainable development has become a fundamental element of state policy in today's Poland. It is included in the most important documents of the country, determines the development strategies at all levels of public administration, and, as a result, has become a vital keyword (quite frequently abused). In national policy, sustainable development has been commonly interpreted as equal opportunities across all regions, in terms of intergenerational equity, territorial equity, and access to innovation – and is directly linked to the Europe 2020 strategy (Ministry of Administration and Digitization 2013).

This emphasis on sustainable development has been twinned with a focus on regional development or, in other words, the development of "provincial Poland." The expression "provincial Poland" carries a variety of meanings. The simplest one refers to the inhabitants of small and medium-sized towns. Statistical classification in Poland divides cities and towns by population into *big cities* (over one hundred thousand inhabitants), *medium-sized towns* (between twenty thousand and one hundred thousand inhabitants), and *small towns* (below twenty thousand inhabitants). This division, although widely used, raises some doubts due to the obvious heterogeneity of cities and towns, and particularly of those classified as medium-sized. Therefore, the statistical division includes additional subcategories to the basic three types, which are used for in-depth analyses (especially in social analyses). However, the three population categories remain the basic criterion for the classification of Polish towns and cities.

Undoubtedly, the emphasis placed on regional development at the time of Poland's EU accession has had a positive effect on the development of towns and villages in "provincial Poland." One can even propose that the results of these efforts are much more visible in small and medium-sized towns than in big cities. It is in small and medium-sized towns where the development of infrastructure, internal migration trends, and the development of civil society are most apparent. Living in the provinces now makes people feel more proud than ashamed. It often means a better quality of life and local patriotism. Life in the provinces is becoming a great honour, giving local residents the opportunity to influence their immediate environment and take greater responsibility for their actions.

This is confirmed by a number of statistical analyses (e.g. GUS 2011, 2012) and also more detailed studies such as Piotr Raźniak's 2000–2010 analysis of the Cracow Metropolitan Area, which not only demonstrated that out-migration was taking place but also reflected upon the socioeconomic processes that caused this phenomenon. Raźniak (2012) showed that, initially, out-migration involved settling in towns adjacent to the large city of Cracow; however, in 2009–2010, the population started to migrate much further away. Great importance was placed on "the desire to live in a more friendly environment in terms of nature, which also offers more peace and quiet" (Raźniak 2012, 70) and the fact that the smaller towns were becoming increasingly wealthy. These factors were complemented by increased investment in public transport, which improved the quality of roads and local infrastructure, and better conditions for cultural activities through the development of public cultural institutions as well as a growing number of private initiatives. Recent years have also brought participatory budgets, in which residents vote to express their opinions on which things part of the public budget should be spent. Altogether, these factors have significantly improved the quality of local life and attracted new inhabitants to smaller cities and towns.

This chapter examines the implementation of the broad concept of *sustainable development* in small and medium-sized towns in Poland with a specific focus on cultural development and the positioning of culture in these local sustainability strategies. The cultural strategies are derived from the general state policy, with sustainable development featured in both local and national policy documents as the main objective. In the context of sustainability, culture is linked primarily to an improved quality of life. Documents about culture (e.g. *National Strategy for Cultural Development 2004–2013* [Ministry of Culture 2004]) put greater emphasis on the sustainable development of culture itself, while local documents usually present both approaches at the same time.

The examples presented in this chapter were chosen because of their different approaches to the place of culture in sustainable development, sometimes irrespective of statements in strategy documents. Through these cases, one may observe an approach in which culture supports the execution of tasks aimed at the sustainable development of a city or region but, at the same time, is focused on operations intended mainly to boost economic growth. The analysis was based on public documents, scientific studies, and statistics, and it presents an array of approaches evident in recent times. To provide some background for this analysis, the chapter briefly outlines the place of small and medium-sized towns in the context of historical development in Poland, and the changes that occurred after the political transformation of 1989 and the EU accession of 2004. In closing, the chapter comments on the effects of implementing these strategies of development and cultural development, and their impact on the improvement of infrastructure, internal migration trends, and the development of civil society.

Small and medium-sized towns in the context of historic development

For several decades during the Socialist period and closely following it, state policy was far from sustainable development (Rychard 2007; Bański 2008; Heffner 2008). Provincial regions were dependent on the ideas of subsequent ruling parties and the investments made to suit the personal interests of those in power, and were frequently the targets of political rewards or punishments. The national government would concentrate industry or place state institutions in selected areas (thus improving the quality of life in those locations), and could centrally control the migration of people through resettlement and work orders. As a reward, it was possible for certain settlements to house many cultural institutions, receive funding for their development, and obtain the status of a *city*, defined as a regional centre of government. However, it was also possible to lose this privilege, which meant a reduced budget for cultural activities and increased censorship.

The decade of the 1990s was a "crash course" in the free market economy. At the beginning of the decade, the population of small and medium-sized towns began to age rapidly as the young moved en masse to the big cities, where the prospects of finding a job and improving their life quality were much better (this, in turn, influenced the development of urban agglomerations). Already in 1990, all major cities had recorded a clear increase in their population (GUS 2011).

Since the beginning of the 2000s, however, a reverse trend has been observed. For example, the city of Łódź had a population of 848,000 in 1990, but it had dropped to 725,000 by 2011 (GUS 2011). Similar population declines, although not as substantial, could be observed in all the major cities in Poland, with two exceptions: Warsaw (the capital) and Cracow, where the population remains constant. In total, Poland had 908 cities in 2011, inhabited by almost twenty-four million people, two-thirds of the country's total population of about thirty-eight million. In 2011, more than half of the urban population (53.4 percent) lived in small and medium-sized towns, while the remaining 45.6 percent lived in the larger cities. Reasons for the shift towards smaller urban communities during this decade include demographic variables, emigration, and internal migration (GUS 2012).

The population shift should also be viewed in the context of two other related trends. First, these migrations are of two types: part of the migration is relatively close and, as a result, leads to the development of urban agglomerations, while the other part entails a complete break from the big cities and very often involves changing one's career and a way of life. Second, an anti-globalisation/pro-local approach has become a popular trend in Poland. This "choice" can be seen as related to state policy aimed at regional development, including the improvement of the quality of life in the provinces and the implementation of sustainable development policies. Development through culture is a crucial element of all policies.

Overall, the changes in Poland in recent years can be briefly described as four-fold: redefining the role and place of the provinces in the spatial structure of the country, searching for one's own identity, consolidating forces to achieve common goals, and learning how to build together. Undoubtedly, these changes were influenced by the Polish accession to the European Union in 2004, the creation of a new system of state governance based on the Treaty of Accession, and the need to adapt Poland's internal legislation to EU law.

The policy of sustainable development in Poland, including the sustainable development of culture

Polish accession to the European Union made it necessary to examine the way Poland functioned and to draw up a comprehensive programme of development. According to the legal acts adopted in Poland, a development policy is a series of actions taken to ensure the sustainable development of the country. As mentioned in the introduction of this chapter, the concept of sustainable development has become an essential element of state policy, included in the most important state documents, determining development strategies at all levels of public administration, and present in all areas of the country's operation. The key national policy, *Poland 2030: The Third Wave of Modernity*, defines sustainable development as "solidarity between regions," which is interpreted to mean equal opportunities – intergenerational, territorial, and in terms of innovation (Ministry of Administration and Digitization 2013, 25).

This document includes stipulations about the role of culture in the implementation of sustainable development policy, which is generally understood as an increase in civic initiatives, cultural activity, and participation in culture – and the strengthening of the creative potential of Poland. In this context, "participation in culture" is described as a pillar of sustainable development and a crucial element to improve the quality of life and people's well-being (Ministry of Administration and Digitization 2013, 42).

In the past few years of implementing this comprehensive programme of development, the ideas of *sustainable development* have penetrated into all spheres of life, and the concept itself has become a popular keyword. This is not an all-talk-no-action situation: strategic policies are supported by specific actions, instruments, and tools for their implementation. Furthermore, the regional strategies are gradually being complemented by additional programmes. For example, in recent years, local development strategies, created by both large and small towns, have been mushrooming all over the country. Just like the national documents, the local strategies also include ideas about sustainable development and the role of culture in the implementation of sustainable development through coherent policies and programmes. The scale of this phenomenon can be seen in more detail through examining the contemporary administrative division of Poland.

Poland is divided into sixteen regions (provinces or *voivodeships*), further subdivided into smaller units called districts (in Polish: *powiats*), with local centres and more than one town in this territory (according to the statistical classification, there are currently 314 *powiats*), and, finally, communes (in Polish: *gminas*), where the local centre often has the status of a town (but not always) and is surrounded by rural villages. Large and medium-sized cities also have commune status (the Statistical Office has recorded 2,479 communes). The emerging development strategies based on administrative units are generally focused on the development of more than just one town or commune. In addition, there is an increasing number of strategies prepared by individual towns, but they are typically conceived for a wider environment (i.e. not just for the individual town itself) and with a wider scope of impact – the plan is to realize the strategy in other nearby towns too. A similar trend occurs in the development strategies of large cities, which try to cater to the needs of their current residents as well as those people living in the broader urban agglomeration.

A second type of development strategy focuses on different functional areas of the state, such as strategies for the sustainable development of culture. According to the general policy document, *Poland 2030: The Third Wave of Modernity*, all policy documents must include provisions concerning regulations about sustainable development. Two of the most important documents of this type are *National Strategy for Cultural Development 2004–2013* (Ministry of Culture 2004) and *Supplement to the National Strategy for Cultural Development 2004–2020* (Ministry of Culture 2005). Their main strategic objective is to achieve sustainable development of culture in the regions, and they place an emphasis on providing appropriate conditions for the development of culture and on creating instruments that will make that possible.

In recent years, at the administrative level, there are more and more documents under the common name of "cultural development strategies." These documents refer to the national cultural strategy (Ministry of Culture 2004) and the general strategic documents for sustainable development but, to a large extent, result from an evaluation of the state's cultural facilities and determine prospects for development. There are clear echoes of a new way of thinking about culture and treating it as a pillar of sustainable development. Local strategies for cultural development are primarily prepared by local cultural institutions, which is the case for small towns such as Sompolno, Rzeczenica, Wołomin, and Szklarska Poręba. To a large extent, a diagnosis of the cultural situation takes into account the role each local cultural institution plays in the implementation of the town's statutory obligation to undertake cultural activities. The most frequently raised issue is the role of culture in sustainable development, which tends to focus in three main areas: in initiatives aimed at stimulating residents into action, in maintaining traditions, and in creating tourism products and influencing the town's image and local labour market. Small and medium-sized towns are increasingly recognising that joint cultural activities contribute to strong local communities, bringing their inhabitants together, giving residents a greater sense of belonging, and making them want to actively participate in their town's development (Plebańczyk 2012; see also Kangas and Sokka, in this volume).

Culture in local sustainable development strategies: two approaches

The examples of Zator and the Cittaslow Network illustrate the two most common, but very different, approaches to the issue of defining roles for culture in local sustainable development. In the first case, the strategy is based on creating a comprehensive tourism product – virtually all activities are focused on the development of a tourism product and making the town richer, and culture is specifically understood as a pillar of sustainable development. In the second case, the focus is on building a network where culture is an important part of the idea of local residents' well-being.

A *tourism-based strategy – Zator*

The town of Zator is located in southern Poland between two large cities – Cracow and Katowice – and among several smaller ones. The town's history goes back to the fifteenth century, when the territorial division of the Duchy of Oświęcim gave rise to the Duchy of Zator, whose seat received the status of a town. The Duchy's legacy includes some pieces of tangible heritage, such as a church, a palace, and the historic urban layout. The monuments themselves, however, have been repeatedly rebuilt and do not constitute a tourist attraction as such. There are many similar places in this part of Poland, and the town does not dwell on these historic features for attracting visitors. In fact, the palace is not accessible to tourists but is used as the headquarters of one of the local companies.

For many years, Zator was home to aggregate mines, which degraded the immediate environment. A significant number of excavations, which were subsequently flooded with water, occupy about 22 percent of the commune's territory. Along with the nearby Skawa and Vistula Rivers, the excavated areas quickly became the habitat of numerous waterfowl. This led to the creation of a nature reserve. In 2005, the town of Zator launched a local sustainable development strategy for the 2005 to 2013 period (Gmina Zator 2005), followed by one for the 2014 to 2022 period (Gmina Zator 2014). The town's strategy focuses on creating the right conditions for business development and improving the quality of life of residents. It entails specific objectives to be achieved and instruments for their implementation.

With the 2005 strategy, Zator decided to rehabilitate the degraded green areas and reintroduce fish farming, including carp, which is a fish traditionally eaten in Poland, especially during the Christmas season. The strategic initiatives primarily included the development of water bodies, the improvement of infrastructure around them, and a number of actions to protect the landscape and wildlife. The strategy included, for example, the creation of educational trails and the Folk Sculpture Gallery, including the area in the "Natura 2000" European network and the European Greenways bicycle trail EuroVelo R4: Cracow – Morawy – Wiedeń. The Dolina Karpia (Carp Valley) covers the area of the historic carp-farming region from the Middle Ages. The entire project, initially aimed at the development of fish processing, slowly evolved towards tourism associated with carp.

The Carp Valley Regional Cultural Centre in Zator, which emerged from a simple cultural centre, has been implementing projects aimed at supporting this objective and now offers an array of activities, including the Carp Festival, angling competitions for children and teenagers, the Best Water Reservoir Manager Contest, the National Canoeing Trip, the Commune's Summer Football Championship, the Winter Volleyball Championship, and the Commune's Harvest Festival. In addition to such activities directly promoting the Carp Valley, the cultural centre also engages in "normal" activities aimed at cultural education and boasts that in recent years it has managed to encourage the residents to participate in its offerings more often.

The town intensely motivates its residents into action – it helps develop small businesses and create tourist infrastructure, improves the quality of life through the development of public and housing infrastructure, and supports the growth of sport and culture. Many initiatives are aimed at building the brand of Zator and at boosting local patriotism. Today, Zator is famous for its regional dish, Zator carp, and its wicker industry. Numerous initiatives are also undertaken by the Society of the Zator Region Enthusiasts. The town is implementing the Leader+ programme (a European Union initiative to support rural development projects initiated at the local level in order to revitalise rural areas and create jobs) through supporting the Local Action Group named the Carp Valley Association. The main objective of the association (identical to the town's strategy) is to support the sustainable development of the Carp Valley, stimulating residents into action and involving them in the implementation of specific projects aimed at local development. Entrepreneurs who are the most active in promoting the town receive the Golden Scale Award, and the most supportive individuals from outside the town are given honorary citizenship.

Through the development of the Carp Valley, the town became so popular that it decided to look for other ideas to improve the quality of life of inhabitants and to strengthen the existing tourism product. In 2008, a private investor, in cooperation with the commune and the Institute of Zoology of Jagiellonian University, launched Zator Land – a huge edutainment park about dinosaurs. Although some criticise the theme park and claim it is at odds with the idea of sustainable development in the context of promoting Zator's culture, history, and natural assets, it should be noted that the place has significantly improved local tourism and inspired other towns to attempt similar initiatives (e.g. a similar amusement park was created in the nearby town of Inwałd).

All of these activities aim to support the town's development, although there are some doubts about whether the one-sidedness of this tourism-focused approach is appropriate. If all activities are concentrated on a single theme, this does not constitute the sustainable development of a rich local culture, and there are no other cultural initiatives apart from those endorsed by the cultural centre. But the town has become recognisable, and the quality of life has improved substantially. The initiatives taken have not only made Zator more popular, but they have also stimulated the local population to become more active. It is worth noting that the local cultural centre hired new staff and improved its infrastructure. It has also significantly expanded its offer, even if it is slightly one-sided towards tourism.

Networking: Polish towns in the Cittaslow Network

A different type of initiative worth mentioning in the context of sustainable development involves the more than ten Polish towns that are members of the Cittaslow International Network. The idea of *slow* has permeated many areas of life. In general, it implies joint efforts of inhabitants, focuses on transparency and openness, and promotes the development of towns sensitive to the needs of hard-pressed and overworked men and women of the twenty-first century. Today, the concept of a *slow life* is becoming increasingly popular and is positioned in contrast to mass culture with its repetitiveness of lifestyles and models.

The Polish National Cittaslow Network was established in 2005. It is an initiative of the regional government of Warmia and Masuria, undertaken as part of the Sustainable Development Strategy of Poland. This region is often referred to as the "green lungs of Poland" or "the land of one thousand lakes." Its relatively small distance from the capital and its rich natural and cultural assets have made Warmia and Masuria the primary tourist destination for the residents of the largest Polish city, Warsaw – a perfect "getaway" from the urban hustle and bustle. This region is a very good example of the exodus of the population to smaller towns, which stems from a desire to improve one's quality of life and to escape from the noisy and overcrowded metropolises. As a consequence, the social composition of residents is changing, and arriving residents are constantly searching for ideas for living in their new environment. Given this context, it was relatively easy to obtain the support of residents to join the Cittaslow Network. Life in the spirit of *slowness* continues to attract new incoming residents.

Members of the Polish network include Lidzbark Warmiński, Biskupiec, Bisztynek, Lubawa, Nowe Miasto Lubawskie, Olsztynek, Reszel, Ryn, Gołdap, Olsztyn, Barczewo, and Dobre Miasto, all located in the region, and one town from outside the region: Murowana Goślina, situated in the western part of Poland, near Poznań. In 2013, resolutions about joining the Network were passed in two other towns outside the region, and they will soon also become members. These towns are marked by rich histories, and their strategic goals are focused on sustainable development based on the idea of *slow life*. What is essential here is using the existing historical, cultural, and natural resources to attain their goals.

The natural direction of the towns' strategic actions is towards the sustainable development of local culture. These towns do not seek to develop mass tourism but cherish the already existing cultural heritage and support numerous cultural initiatives, which are mainly aimed at improving the quality of life of town dwellers. Initiatives in force in all the towns belonging to the Network include conducting an annual inventory of traditional products; organising local cultural events; and supporting initiatives that encourage the protection of regional products and crafts.

All the towns have long and rich histories, where cultural monuments are the tangible legacies. For example, in Lidzbark Warmiński one can admire the fourteenth-century castle of the bishops of Warmia; the town of Reszel features a medieval urban layout from the nineteenth century; and Gołdap, entirely ravaged during the world wars, was rebuilt as a health resort. The methods for improving the quality of urban life in the Network's towns have included specific actions such as restoring historical centres, revitalizing objects of cultural or historical value, and developing programmes to improve the status and accessibility of historical centres.

The incoming residents, mostly the ex-residents of big cities, are the added value. They pursue their passions, often in relation to the traditions of the place where they decided to settle. They tend to blend into the local communities, restoring traditional professions and motivating others into action. In this way, more regional products appear, which, on one hand, stimulate social and economic development, and, on the other hand, become tourist products themselves.

Culture-led sustainable development: other approaches

What is typical of today's small and medium-sized Polish towns is that they seek to be different and unique, wanting to find something that will make them stand out from the crowd. In order to implement a policy of sustainable development, it is necessary to reflect upon the available resources and their use. By doing so, towns sometimes discover their cultural resources anew, such as derelict historic monuments, local cultural centres, or records about local history. In this way, it is possible to distinguish one more element that is distinctive in the processes of implementing sustainable development policy in many small and medium-sized towns, which could be called *culture-led sustainable development*.

The majority of Polish towns and cities have long-standing traditions and rich histories to which they may refer, and they are doing it more and more often. This is not only the case for big cities, however. The little town of Sejny is a very distinctive example. It has become world-famous thanks to the Centre "Borderland of Arts, Cultures and Nations" (the Borderland Foundation). The town is located on the eastern border of the country where, for hundreds of years, people of different nationalities and religions have lived, and where there are towns with Catholic churches, Jewish synagogues, Orthodox churches, and Muslim mosques standing next to each other. Culture or, rather, cultural diversity is the basis of existence for Borderland. Culture here is a source of development and a development goal. Borderland undertakes many initiatives supporting the development of culture, such as the Summer School of Intercultural Dialogue, New Agora, and the Borderland School, open to people from the whole world, and other initiatives like the Sejny Chronicles, Atlantis Borderland (a cross-border cultural trail), and school workshops, which are dedicated more to local residents. It helps encourage and support intercultural dialogue between national minorities and ethnic groups, and it has a very strong impact for the local population and their awareness of their own history and the history of the place where they are living. The best example is a history of Sejny's Jews, who did not return there after World War II, but left behind both tangible and intangible heritage. The Borderland Centre is well known abroad and attracts volunteers from around the world.

A second example involves the cities of Silesia, a region that, for many years, has been used almost exclusively for coal mining. The closing of mines not only led to unemployment but also triggered a problem regarding the development of post-mine facilities. Paradoxically, Silesia, perceived to be a predominantly industrial region, decided to invest in the development of culture. Some towns and cities turned their mines into very attractive museums that present the history of mining in the area and often the heritage more generally (e.g. the town of Tarnowskie Góry). Through such efforts, a very attractive tourism product has emerged. However, the initiatives are not only tourism focused. The towns and cities have also taken a number of actions to improve the quality of their residents' lives. For example, some former mines (as well as other industrial facilities, such as factories) are not only museums but also small cultural centres that are home to theatrical performances and concerts and have a wide offer of activities for local residents.

"Poland under construction" – this widespread slogan, created by the media, accurately illustrates recent years in Poland, referencing its new road infrastructure, the development of transport, and new public buildings. The best-known cultural infrastructure investments are found in large cities, where the largest projects are implemented, like the construction of new major facilities for cultural activities. (Investments in the modernisation of local libraries or community centres have been less noticeable.) Nevertheless, some of the most spectacular cultural infrastructure investments have been located in medium-sized towns, for example, the construction of Podlasie Opera and Philharmonic – European Centre for the Arts in Białystok, and the establishment of the Centre for Artistic Education – Philharmonic Hall in Gorzów.

Revitalising historic buildings that highlight the specificity of the region and using them for cultural purposes, as well as developing new cultural and recreational facilities, raises the attractiveness of towns and regions by increasing the influx of tourists and improving the quality of life. Such investments contribute to increasing a town's

- residential appeal – the quality of public space is higher, more highly skilled workers move to towns, and more attractive jobs are created;
- economic appeal – new jobs are gained, demand for cultural goods and services increases, and the economic structure of towns can change;
- tourist appeal – tourist products are created; and
- lifestyle appeal – residents gain improved access to cultural facilities and offerings.

(Wojnar and Grochowski 2011)

In addition, in the cases examined for this chapter, all investments have had the added value of strengthening residents' sense of local identity and pride and boosting their self-esteems.

Analysis of internal migration in Poland has clearly highlighted a trend of abandoning big cities and either moving to the suburbs or relinquishing big city life altogether and trying to make it in small towns. Exoduses of the population to smaller towns are triggered by many factors, including the desire to improve the standard of living, to escape from the urban hustle and bustle, and to say "no" to globalisation. Distance from a larger town has ceased to be an inhibiting factor – the need to find a job or the desire to access goods and services not available in one's current place of residence is now less of an issue, and this has significantly increased population mobility.

Small and medium-sized towns have a growing appeal. Their inhabitants see quick and tangible results of their efforts, which tend to motivate them into taking further actions. They feel more and more responsible for the way their local community functions, and this tends to foster better integration, joint management of the town, and real sustainable development. All this makes the prospect of living "in the provinces" increasingly attractive. The result is a changing social cross-section of small and medium-sized town residents. Newcomers often become a stimulus for the development of local communities; they blend in but also activate existing local residents.

Culture plays a very important role here. Almost all local development strategies contain provisions about the sustainable development of culture, which derives from national documents and a legal obligation to provide cultural activities in a city or a town. Sometimes, it merely boils down to having a local cultural institution; however, more often it means providing the conditions for cultural development and being open to new initiatives. This is best illustrated by a phenomenon especially popular in Polish villages – open-air events (festivals) that present a very wide range of local cultural activities from folk, theatre, and music through the works of local artists and groups, to local cuisine. It can be observed that the cultural activities offered by such events unite different generations and are a driving force behind small businesses. Through this, culture becomes a real factor for improving the quality of life and a pillar of sustainable development.

References

Bański, J. (2008) Przemiany funkcjonalno-przestrzenne terenów wiejskich – diagnoza, rekomendacje dla KZPK [Functional and spatial transformation of rural areas – diagnosis, recommendations for NSDC]. In *Ekspertyzy do Koncepcji Przestrzennego Zagospodarowania Kraju 2008–2033 [Expert Evaluations of the National Spatial Development Concept 2008–2033]* (pp. 397–432). Warsaw: Ministerstwo Infrastruktury i Rozwoju. Available at: www.mrr.gov.pl/rozwoj_regionalny/poziom_krajowy/polska_polityka_przestrzenna/prace_nad_KPZK_2008_2033/Documents/Banski.pdf [Accessed 12 November 2014].

Główny Urząd Statystyczny (GUS) (Central Statistical Office of Poland) (2011) *Migracje Wewnętrzne Ludności* [*Internal Migrations of the Population*]. Available at: www.stat.gov.pl [Accessed 18 November 2014].

Główny Urząd Statystyczny (GUS) (Central Statistical Office of Poland) (2012) *Wyniki Narodowego Spisu Powszechnego Ludności i Mieszkań 2011* [*The Results of the National Census of Population and Housing 2011*]. Available at: www.stat.gov.pl [Accessed 18 November 2014].

Gmina Zator (2005) *Strategia Rozwoju Gminy Zator 2005–2013* [*Strategy for the Development of the Zator Commune 2005–2013*]. Available at: www.zator.pl [Accessed 10 November 2014].

Gmina Zator (2014) *Strategia Rozwoju Gminy Zator 2014–2022* [*Strategy for the Development of the Zator Commune 2014–2022*]. Available at: www.zator.pl [Accessed 19 November 2014].

Heffner, K. (2008) Funkcjonowanie miast małych w systemie osadniczym Polski w perspektywie 2033 – rekomendacje dla KPZK [Functioning of small towns in the Polish settlement system in the year 2033 – recommendations for NSDC]. In *Ekspertyzy do Koncepcji Przestrzennego Zagospodarowania Kraju 2008–2033* [*Expert Evaluations of the National Spatial Development Concept 2008–2033*] (pp. 285–333). Warsaw: Ministerstwo Infrastruktury i Rozwoju. Available at: www.mrr.gov.pl/rozwoj_regionalny/poziom_krajowy/polska_polityka_przestrzenna/prace_nad_KPZK_2008_2033/Documents/Heffner.pdf [Accessed 12 November 2014].

Ministerstwo Administracji i Cyfryzacji (Ministry of Administration and Digitization) (2013) *Polska 2030: Trzecia fala nowoczesności: Długookresowa Strategia Rozwoju Kraju* [*Poland 2030: The Third Wave of Modernity: Long-Term Development Strategy of the Country*]. Available at: https://mac.gov.pl/files/wp-content/uploads/2013/02/Strategia-DSRK-PL2030-RM.pdf [Accessed 20 October 2014].

Ministerstwo Kultury (Ministry of Culture) (2004) *Narodowa strategia rozwoju kultury na lata 2004–2013* [*National Strategy for Cultural Development 2004–2013*]. Available at: http://bip.mkidn.gov.pl/media/docs/Narodowa_Strategia_Rozwoju_Kultury.pdf [Accessed 20 October 2014].

Ministerstwo Kultury (Ministry of Culture) (2005) *Uzupełnienie narodowej strategii rozwoju kultury na lata 2004–2020* [*Supplement to the National Strategy for Cultural Development 2004–2020*]. Available at: http://bip.mkidn.gov.pl/media/docs/050617nsrk-uzupelnienie.pdf [Accessed 20 October 2014].

Plebańczyk, K. (2012) Civil society, participation, and urban games – comments on the research in the context of culture management. *Culture Management*, 5(1), 15 pp. Available at: http://culturemanagement.pl/?p=55 [Accessed 5 November 2014].

Raźniak P. (2012) Procesy społeczno-ekonomiczne w Krakowskim Obszarze Metropolitalnym [Socio-economic processes in the Cracow Metropolitan Area]. *Prace Geograficzne*, 129, 63–81.

Rychard, A. (2007) Socjologiczne uwarunkowania rozwoju przestrzennego Polski – rekomendacje dla KPZK [Sociological determinants of Poland's spatial development – recommendations for NSDC]. In *Ekspertyzy do Koncepcji Przestrzennego Zagospodarowania Kraju 2008–2033* [*Expert Evaluations of the National Spatial Development Concept 2008–2033*] (pp. 453–467). Warsaw: Ministerstwo Infrastruktury i Rozwoju. Available at: www.mrr.gov.pl/rozwoj_regionalny/poziom_krajowy/polska_polityka_przestrzenna/prace_nad_KPZK_2008_2033/Documents/Rychard.pdf [Accessed 12 November 2014].

Wojnar, K., and Grochowski, M. (2011) *Infrastruktura kultury: Polityka spójności, atrakcyjność miast* [*Infrastructure of Culture: Cohesion Policy, Attractiveness of Towns and Cities*]. Warsaw: Narodowe Centrum Kultury.

12 Culture, sustainable development, and innovation

The case of Norrby, Sweden

Jenny Johannisson

Introduction

> The city district of Norrby is rumored to be a problem area; ugly, poor and dangerous. The average income level is low, the level of unemployment is high and the feeling of insecurity is greater than in the rest of the city. Media tends to pick up and report on negative events and circumstances in Norrby.
>
> More than two years of close studies through Cultural Planning provides a much more nuanced picture of the city district and its inhabitants.
>
> Norrby is centrally located and well united. There are several businesses, a bank, a school, a community youth centre, an open preschool and a community centre with a library. Housing is generally well kept and green areas could be developed into proper oases.
>
> The area is characterized by a diversity of people, languages and knowledge, and the inhabitants are deeply engaged in their own local living environment and city district. The inhabitants, to a great extent, are agreed in their approval of Norrby as a nice place to live. There are problems, but they can be solved!
>
> (City of Borås 2013, 1 [author's transl.])

These conclusions were drawn from the City of Borås' Cultural Planning project. In a Swedish context, it is mainly at the local level of government and in the urban settings of small and medium-sized cities that sustainable development has been translated into actual cultural policy initiatives (Johannisson 2006, 2012). In such initiatives, cultural planning has over the last decades been increasingly introduced as a policy tool that furthers the tripartite aim of sustainable development – that is, to use culture as a means to simultaneously promote economic and social sustainability in ways that minimize negative effects on ecological sustainability. The actual outcomes of cultural planning remain, however, largely unexplored in a Swedish context. In addition, these outcomes need to be more clearly related to the discursive underpinnings of cultural planning. While several observations made concerning these underpinnings in a British context (Connolly 2013; McGuigan 2004; Stevenson 2004; Stevenson, Rowe, and McKay 2010) are also relevant in Swedish, European, and global contexts, a more nuanced understanding of actual cultural planning initiatives implies that such initiatives be explored in the historically and geographically specific settings that surround them, settings that tend to shape global features in ways that are, indeed, local.

In this chapter, I will trace the formation of the Cultural Planning initiative in Norrby, a district of the City of Borås, in western Sweden. The City of Borås is Sweden's thirteenth-largest municipality with 106,000 inhabitants, where approximately 70,000 live in the actual city. Borås clearly belongs to the category of medium-sized European cities, although from a national perspective the city is quite large. Together with 35.5 percent of the population in the European Union, Borås belongs to the category of *intermediate regions*, according to the OECD classification of European regions based on the region's share of population living in "rural local units" (INTELI 2011, 33–34).[1] The city was founded in 1621, and industrialization processes during the nineteenth century established the textile industry as the city's major trademark. Today, the public sector, together with small and medium-sized businesses in the textile and mail-order sectors, employ most people.

Norrby, located right next to the Borås city centre, has approximately 3,500 inhabitants, a majority of whom are immigrants. Like many city districts in Sweden, many of Norrby's older buildings have been demolished, particularly through initiatives taken by the Social Democratic government during the 1960s and 1970s, and replaced with buildings in the architectural style of modernism. In the case of Norrby, the textile factory buildings and worker's housing were torn down, thus obliterating important parts of the city's cultural heritage. In spite of its central location – a characteristic usually represented as an advantage in a Western European context – Norrby has a poor reputation due to the high number of immigrants as well as its high level of unemployment, vandalism, and other criminal activities, attributes which are usually associated with the suburbs of Swedish cities (Sundin 2007). Norrby is represented as a challenge by the municipality of Borås, one that needs to be attended to in different ways. As buildings from the 1960s and 1970s need to be refurbished, and since Norrby includes a strip of land that remains to be developed, the municipality has, for some time, planned to use a combined building/rebuilding project as a strategic arena for meeting these challenges.

In 2012, the City of Borås together with SP Technical Research Institute of Sweden, Borås Energi och Miljö (a local energy company), and the University of Borås instigated a city development project to use the building/rebuilding project in Norrby as a stepping stone for addressing social, ecological, and economic issues in Norrby. The 2013–2015 project, Innovation Platform Norrby, is financed partly by VINNOVA (Swedish governmental agency for innovation systems) and partly by the platform members. It aims to form an innovation platform for sustainable urban futures that involves public-private partnerships and the bringing together of different kinds of competences – political, administrative, industrial, and academic – in achieving such futures. While the building/rebuilding process is the material starting point of the project, culture is explicitly introduced as a tool in involving the citizens in the project and, ultimately, in the shaping of their own urban futures.

The pre-study drawn upon in this chapter is part of this city development project, and it maps the enactment of cultural planning as it is put into practice. The pre-study, and the subsequent full case study that will be carried out during 2014–2015, aims to critically explore the different meanings attributed to culture, sustainability, and innovation by key agents in the project and to trace the cultural policy discourses put to use in the process. My main concern is with cultural policy making and the different ways in which cultural policy positions culture in relation to itself as well as to other policy fields. However, I also want to explore how policy makers in other fields, as well as agents outside the political-administrative system, perceive and communicate culture.

From a primarily neoinstitutionalist perspective on policy making, this chapter aims to provide an analysis of a local expression of the interrelations between culture, sustainable development, and innovation and to make comparisons with other such examples possible, on a European or potentially global scale. Culture, sustainable development, and innovation are understood in terms of three separate "organizational fields" (DiMaggio and Powell 1983) – that is, as "a construct that allows us to conceptualize how organizations concerned with similar issues but from a variety of sectors might form some sort of unified system" (Strom 2003, 251). As Elizabeth Strom (2003, 262) notes, in political science what separates an organizational field from a policy community is scope: while policy communities are "concerned primarily with interactions that are intended to shape policy," organizational fields are concerned with exploring any kind of relation between any kind of agents within the field. Although policy making is the primary focus of this chapter, I find that the case of Innovation Platform Norrby can be more fully explored with this broader understanding of relevant relationships. The organizational fields of culture, sustainable development, and innovation are internally related through common ideas and norms. In Innovation Platform Norrby, the three fields are brought together, generating new relations within and between fields that potentially challenge the internal corpus of ideas and norms within each field. In this chapter, I trace the ideas and norms within each field while gradually introducing the formation of Innovation Platform Norrby, and I provide an initial analysis of potential points of contact and potential points of conflict both within and between the fields, as enacted in the project.

Culture

The agents, values, norms, and initiatives within the field of culture could be identified as one of several sets of important driving forces leading up to Innovation Platform Norrby. In this project, the City of Borås is the key agent. In 2011, the municipal cultural committee instigated a two-year Cultural Planning project in Norrby, aiming at "strengthening democracy, increasing quality of life, and making use of place-generated cultural resources" (City of Borås 2013, 29). Cultural planning, as a method in public cultural policy, has gradually been translated into the Swedish context over the last twenty years, adapting from research-based initiatives mainly in the UK, Australia, Canada, and the US (Stevenson 2004). One of its most prominent advocates, Franco Bianchini, introduces cultural planning as a perspective rather than a specific method:

> In conclusion, a cultural planning perspective, rooted in an understanding of local cultural resources and of cities as cultural entities – as places where people meet, talk, share ideas and desires, and where identities and lifestyles are formed – could help planners assess the needs of the community, ensure cultural pluralism and conceptualise essential strategic questions about the city's future.
>
> (Bianchini 1993, 212)

In this quote, Bianchini also provides important reasons for why cultural planning has had difficulties in making an impact on Swedish cultural policy at the national level as well as at the local level of government. Until the 1980s, in the cultural field of Sweden and the other Nordic countries, public cultural policy has relied on two basic ideas, one translated from the perspective of artists and art organizations and one from the perspective of organizations in civil society and the social democratic movement of the twentieth century. The first idea has generated the norm that public cultural policy should primarily promote the production and dissemination of professional artistic activities of high quality, safeguarding the inherent value of art in both policy making and policy implementation (Mangset et al. 2008). The second idea partly supports the first one, generating the norm that public cultural policy should make professional art of high quality accessible to every citizen since art (as a resource) tends to be unevenly distributed between different social groups. Within this framework, art is important as a public resource because it contributes to the "spiritual" development of the individual citizen and therefore of society as a whole (see Johannisson 2012; McGuigan 2004). In a study of the conflicting positions of urban planners and cultural workers in culture-based urban development strategies in a Norwegian city, Hans Kjetil Lysgård (2013) refers to this set of ideas as the "cultural development discourse" (194). In Lysgård's analysis, cultural workers and artists promote this discourse in opposition to the "urban development discourse" promoted by urban planners and policy makers. Lysgård concludes by tracing one important point of conflict between the two discourses – namely, that of conflicting interpretations of a spatial perspective: "One discourse – the urban development discourse – includes spatiality as a fundamental intention, while the other – the cultural development discourse – has either the individual's development as the focus or the expression of art's inherent quality as a fundamental intention" (Lysgård 2013, 197).

As a policy tool, cultural planning has had problems being integrated into public cultural policy in Sweden because it tries to combine the two discourses that Lysgård identifies as antagonistic. As Bianchini states, cultural planning is about emphasizing place-specific and local aspects of artistic activity, infusing them with a spatial dimension, and turning them into "cultural resources." In cultural planning, the inherent value of art is not questioned, but it is, rather, a nonissue; cultural planning is about using culture as a tool to further primarily social aims generated by the citizens themselves. This perspective is evident in the report from the Cultural Planning project in Norrby, where the objective of using cultural planning is to "develop and make use of the cultural resources of Norrby to further integration" (City of Borås 2013, 29). The list of measures to achieve this goal presents cultural planning as a method, rather than Bianchini's way of describing it as a perspective. As a method, cultural planning is about transgressing borders between different sectors within the public sector, as well as borders between the public sector, on one hand, and private and civil sectors, on the other. It is about marketing and anchoring the process broadly, not least amongst the inhabitants of Norrby, and about "mapping" (Mercer 2002) the cultural resources of Norrby. It is about generating a "sustainable action plan for Norrby, both in a short-term and long-term perspective," and it is about turning Norrby into a "meeting-place" (City of Borås 2013, 29).

The results of the "cultural mapping" of Norrby show that the City of Borås uses these measures consistently. The "cultural resources" listed range from traditional artistic artifacts and art-mediating practices such as sculptures and painting classes for children, to playgrounds and associations based in popular movements. Culture-based urban development strategies have been criticized internationally for allowing a market-oriented discourse to gradually dominate public policy making (McGuigan 2004; Stevenson 2004; Stevenson, Rowe, and McKay 2010). Judging from the report from Norrby, however, it is evident that local cultural policy makers maintain one of the traditional focal points of Swedish cultural policy, that of culture as a tool for increased participation in cultural life, linking cultural policy with issues of social inclusion and social sustainability.

Sustainable development

In the conclusion of the report from the Cultural Planning project at Norrby, the results are described as being in line with the general objectives of the City of Borås. In 2012, the City Council endorsed a vision of the city for 2025 as well as strategies to achieve this vision. In this key policy document, sustainable development is the self-evident starting-point: "We strive for sustainable development which integrates ecological, social and economic dimensions. Awareness of how our habitat is influenced by our choices is great, and motivates us to promote sustainable development" (City of Borås 2012, 5 [author's transl.]).

The politicians of Borås are thus well tuned to the integrative approach to sustainable development that has gradually been shaped and adapted by national and international governmental agencies since the report of the Brundtland Commission in 1987 (Nolin 2010; Throsby 1997). In particular, the City of Borås is manifestly interested in furthering the social dimension of sustainable development, particularly through promoting cooperation and partnerships between different sectors and place-specific agents: "Collaboration between business, university, research centres and public agents is our trademark and the City of Borås pursues new ways of co-operating and thinking" (City of Borås 2012, 6 [author's transl.]). Innovation Platform Norrby could be considered one stepping stone in finding such new ways of integrating different perspectives to reach common objectives. Running parallel with the Cultural Planning project in Norrby, the City of Borås, the University of Borås, and the SP Technical Research Institute of Sweden signed a letter of intent establishing a "Partnership for Sustainable City Development." While this partnership focuses on building techniques, it uses Norrby as a specific case when striving to integrate the more technical aspects with "problems of a more sociological character, for example an ageing population, loneliness, segregation and feelings of insecurity in a broad sense" (City of Borås, University of Borås, and SP 2013, 5 [author's transl.]). In particular, the partnership aims to contribute to meeting the challenge of having over thirty different nationalities in the specific space of Norrby through fostering a higher degree of integration within Norrby and within the City of Borås as a whole. The fact that the undeveloped strip of land in Norrby is also the geographical strip that links Norrby directly to the city centre is a very physical expression of the primarily social project of building/rebuilding Norrby. Starting from the ecological and economic aspects of how to achieve sustainable growth in Norrby and Borås, the partnership as well as the subsequent Innovation Platform Norrby are left with issues that are primarily social in character.

A key aspect of social sustainability is that it is not achieved solely through the wise decisions of public and other agents for the rest of the population. Instead, furthering social sustainability involves the citizens, customers, and users that inhabit a specific place in the planning process. Therefore, surveys and workshops directed to the inhabitants of Norrby are considered a key method in Innovation Platform Norrby. This method is also what seemingly separates cultural planning from previous ideas and norms in Swedish public cultural policy. Such policy has primarily been shaped and implemented by professional agents in the cultural field who refer to professional expertise when deciding what artistic and cultural activities should be supported by public means in order to reach a wider section of the population (Lysgård 2013). Cultural planning turns this approach around, arguing that the inhabitants of a specific place are experts on what should be promoted in their own neighbourhood. Since publically funded professional artistic activities resonate only with a narrow segment of the population, making use of cultural mapping as a tool usually relies on a broad, anthropological conception of culture. From this perspective, artistic activities are included in cultural resources, but the focus tends to be predominantly directed at a great variety of human practices, including the "ways of life" (Williams 1985) of particular groups of inhabitants.

This broad conception of culture is manifestly expressed in the vision for Borås in 2025, which states: "Culture is characterized by both breadth in scope and efforts of particular distinction, and it involves a high degree of citizen participation" (City of Borås 2012, 6 [author's transl.]). The anthropological conception of culture is in the case of the Cultural Planning project in Norrby most explicitly expressed when the Cultural Planning Laboratory, a network of municipalities working with Cultural Planning in Western Sweden, introduces it as a method that helps "finding the soul of a specific place" (Cultural Planning Laboratory 2014). The City of Borås is a member of this network and uses not only the network's wordings on culture in this anthropological sense but also its way of framing Cultural Planning primarily in terms of local development: "Cultural Planning is a method for making local cultural resources visible and integrating them with local development and community planning" (City of Borås 2013, 28 [author's transl.]).The framework of local development is more evident in the report than the framework of sustainable development, which is rather used to relate Cultural Planning activities to the general objectives stated in the vision for Borås in 2025. The report thus establishes Cultural Planning as a tool that is put to work at the more concrete, specific, and detailed level, in efforts to attain the general objectives of public policy making in Borås. Even more specifically, Cultural Planning in Norrby is primarily about furthering integration – between different cultures in the anthropological sense – which is considered an important expression of social sustainability. To achieve this goal, policy making has to become more innovative.

Innovation

The third organizational field involved in the rebuilding/regeneration project in Norrby is that of innovation. Even though innovation is used by the "Partnership for Sustainable City Development" as a way of arguing for the positive output the partnership will generate, it becomes a significant keyword when the constituent members of the partnership, together with the municipally owned local energy company, Borås Energi och Miljö, apply for funding from VINNOVA for the Innovation Platform Norrby. The mission of this national governmental agency for innovation systems is to "promote sustainable growth by improving the conditions for innovations, as well as funding needs-driven research" (VINNOVA 2014), and it is guided by the following vision:

> VINNOVA's vision is for Sweden to be a world-leading country in research and innovation, an attractive place in which to invest and conduct business. We promote collaborations between companies, universities, research institutes and the public sector. We do this by stimulating a greater use of research, by making long-term investments in strong research and innovation milieus and by developing catalytic meeting places.

The primary focus of VINNOVA is to promote sustainable growth, in an economic sense, in the private sector. Innovation is thus first and foremost understood as an ability to transform new ideas and new ways of working into new business opportunities – that is, to develop new products and services to be sold in a competitive market. However, VINNOVA is also paying increasing interest to social innovation in the public sector, which it defines as "a measure which aims at increasing people's well-being by identifying and meeting social needs"

(VINNOVA 2014). When Innovation Platform Norrby was granted funding from VINNOVA, it was mainly with reference to the potential of the platform to support social innovation, even though the platform will also have to document innovations of the more traditional kind. In every meaning of the word, for both VINNOVA and Innovation Platform Norrby, *innovation* is primarily about enabling collaboration between different agents, organizational fields, and ideas to promote new ways of thinking. The most fundamental definition of innovation thus in some way relies on what Henry Chesbrough (2003) has successfully launched as "open innovation." In the memorandums of a meeting between Innovation Platform Norrby and other VINNOVA-funded platforms on sustainable city development, VINNOVA defines *open innovation* as the following:

> to systematically pursue innovations external to your own organization, and to invite several actors, large and small, to participate in the innovation process. Open and transparent innovation processes can be found in the public sector, the private sector and amongst consumers/users and they are often distinguished by being collaborative and driven by needs.
>
> (Memorandums 2014, 1 [author's transl.])

Open innovation is one of the key words in European Union policy making, especially for research and innovation policy in the Information and Communication Technology area. Agents within this area have developed a specific method, Living Labs, generated from the perspective of open innovation and increasingly gaining influence in Europe since the year 2000 (Ståhlbröst 2008). In a report that documents the use of this method in several European countries, the European Commission provides the following definition:

> Living Labs are open innovation environments in real-life settings, in which user-driven innovation is fully integrated within the co-creation process of new services, products and societal infrastructures. In recent years, Living Labs have become a powerful instrument for effectively involving the user at all stages of the research, development and innovation process, thereby contributing to European competitiveness and growth.
>
> (European Commission 2009, 5)

The perspective of open innovation and the method of Living Labs are thus basically about involving users as active participants in research and innovation processes, mainly related to some kind of technology. In the Innovation Platform Norrby, the single method of Living Labs is used by all six work packages in the platform – that is, as the method of the innovation platform itself. The method was introduced to the platform by SP Technical Research Institute of Sweden, owned by the Swedish national government, which positions itself as an important link between research, innovation, and commercialization.

The method of Living Labs clearly has similarities with the method of cultural planning, but it is also obvious that they are generated from different organizational fields, which are governed by different ideas and norms upheld by different agents. As Lysgård (2013) shows in his study on culture-based urban development strategies, innovation is understood very differently in what he labels the cultural development discourse and the urban development discourse. While the first discourse "links innovation to an individual's learning, education and self-realisation," a gradual process that infuses the discourse with a hierarchical perspective where professionalism is central, the latter is concerned with "social improvement and economic growth" and in broadening both the very concept of culture and the scope of agents that are to be included in shaping it (Lysgård 2013, 194–195). With the terminology used here, the organizational field of culture, on one hand, and the organizational fields of sustainable development and innovation, on the other, are infused with the same, potentially contradictory, ideas and norms concerning innovation. Perhaps most importantly, the dominant understanding of innovation is in the platform translated from the organizational field of innovation, where culture is primarily understood as organizational culture – that is, as a specific way of thinking and working that promotes common objectives by transgressing borders between different fields.

Concluding remarks

In this chapter, I have traced the discursive underpinnings of Innovation Platform Norrby, a project that aims at rebuilding/regenerating a city district in a sustainable way – ecologically, financially, and socially. It is about bringing together parallel processes in cultural policy making, urban planning, and research-led innovation to further a mutual goal. I have shown that culture, in the explicit sense that denotes specific artistic artifacts and practices as well as cultural practices understood more broadly, is introduced to the platform by way of cultural planning, and mainly by cultural policy agents in the City of Borås. The anthropological understanding of culture that primarily underlies cultural planning is also at the fore in the platform's understanding of sustainable development. Culture is first and foremost considered as a tool that can foster social sustainability by way of fostering integration between the many ethnic groups living in Norrby, on one hand, and the Swedish ethnic group, on the other. Urban planners and general policy makers in the City of Borås are, together with the University of Borås, the main driving forces behind this understanding. Finally, specific organizational cultures that enable user-oriented innovation infuse the platform with a perspective that is basically technological and market-oriented in character. Here, general policy makers in the City of Borås join forces with VINNOVA, SP Technical Research Institute, and Borås Energi och Miljö, the local energy company. Even though the analysis presented here is only based on a pre-study, it seems evident that the platform – in aspiring to merge ideas and norms from the very different organizational fields of culture, sustainable development, and innovation – is both unique and an expression of more global trends in cultural policy and urban planning.

It is unique in the sense that it is based on highly specific historic and geographic local conditions and agents, infusing the discussions on culture and sustainable development with a truly spatial perspective that is often missing in the context of Swedish public cultural policy making. It is not merely a truism to state that Innovation Platform Norrby could not have been accomplished elsewhere, since the local conditions and agents of Borås and Norrby determine its shape and future trajectory. Concerning the aim and working methods of the platform, however, Innovation Platform Norrby points to more general traits in at least two different, though highly interlinked, ways. First, the processes that shape and uphold the platform clearly express that local settings are never shaped, upheld, or transformed in isolation from other spatial scales; in establishing the platform, the agents involved relate to other processes in other local, regional, national, and international settings (Johannisson 2010). Indeed, the platform is very explicit in wanting to link local settings with other settings on a global scale. Secondly, these other processes that Innovation Platform Norrby is both shaped by and contributes to shaping seem increasingly global in character. As Deborah Stevenson (2004) has shown in a highly critical article on its assumptions and consequences, cultural planning is an explicit expression of such processes. According to Stevenson, it is not even evident that cultural planning actually brings about the positive results that it purports to achieve. Instead, cultural planning has primarily resulted in interlinking the cultural, social, and economic through tautological logics that "disguise significant political motives and assumptions" (Stevenson 2004, 128).

A case in point is the understanding of social inclusion that is translated into Innovation Platform Norrby mainly from the fields of sustainable development and innovation. Social inclusion is, as Stevenson (2004) observes, something quite different from social justice; the former is about the opportunity to "'participate' in society" (26), while the latter is about access for all social groups to social, economic, and cultural resources. From a policy-making perspective, this is a crucial point since "the goal of social justice requires an interventionist state with a redistributive agenda, while social inclusion legitimizes mutual obligation and 'small' government" (26).

Stevenson's criticism of cultural planning is thus relevant also to Innovation Platform Norrby. In bringing together ideas and norms from different organizational fields – in this case, under the umbrellas of cultural planning, sustainable development, and innovation – there is always a risk that the inherent values of each field are lost. A full assessment of the results in Norrby is not yet possible, but it would seem that while the platform has already resulted in the recognition of the value of culture by agents external to the field of culture, it is not the ideas and norms of the cultural field that are allowed to dominate the platform.

Note

1 In a national context, most of western and southern Sweden (as well at the far north) belongs to this category, while the majority of Swedish territory is classified as predominately rural. The capital region of Stockholm singles itself out completely in being classified as predominately urban (INTELI 2011, 33).

References

Bianchini, F. (1993) Culture, conflict and cities: Issues and prospects for the 1990s. In F. Bianchini and M. Parkinson (eds.), *Cultural Policy and Urban Regeneration: The West European Experience* (pp. 199–213). Manchester: Manchester University Press.

Chesbrough, H. W. (2003) *Open Innovation: The New Imperative for Creating and Profiting from Technology*. Boston: Harvard Business School Press.

City of Borås (2012) *Borås 2025: Vision och strategi* [*Borås 2025: Vision and Strategy*]. Borås: City of Borås.

City of Borås (2013) *Stadsdelen Norrby 2011–2013: Cultural Planning i Borås stad* [*The City District of Norrby 2011–2013: Cultural Planning in the City of Borås*]. Borås: City of Borås.

City of Borås, University of Borås and SP (2013) *Partnerskap för hållbar stadsutveckling* [*Partnership for Sustainable City Development*]. Borås: City of Borås.

Connolly, M. G. (2013) The Liverpool model(s): Cultural planning, Liverpool and Capital of Culture 2008. *International Journal of Cultural Policy*, 19(2), 162–181.

Cultural Planning Laboratory (2014) Available at: http://culturalplanninglaboratory.se/ [Accessed 28 February 2014].

DiMaggio, P.J., and Powell, W.W. (1983) The iron cage revisited: Institutional isomorphism and collective rationality in organizational fields. *American Sociological Review*, 48, 147–160.

European Commission (2009) *Living Labs for User-Driven Open Innovation: An Overview of the Living Labs Methodology, Activities and Achievements*. European Commission, Information Society and Media, doi:10.2759/34481.

INTELI (2011) *Creative-Based Cities in Small and Medium-Sized Cities: Guidelines for Local Authorities*. Lisbon: INTELI. Available at: http://urbact.eu/fileadmin/Projects/Creative_Clusters/documents_media/URBACTCreativeClusters_TAP_INTELI_Final.pdf [Accessed 10 March 2014].

Johannisson, J. (2006) *Det lokala möter världen: Kulturpolitiskt förändringsarbete i 1990-talets Göteborg* [*The Local Meets the World: Cultural Policy (Re)construction in the City of Borås during the 1990s*]. Borås: Valfrid.

Johannisson, J. (2010) Making geography matter in cultural policy research: The case of regional cultural policy in Sweden. In J. P. Singh (ed.), *International Cultural Policies and Power* (pp. 127–139). Basingstoke: Palgrave MacMillan.

Johannisson, J. (2012) Sustainable development and regionalization in Swedish cultural policy. Paper presented at the 7th International Conference on Cultural Policy Research (ICCPR 2012), Barcelona, 9–12 July 2012.

Lysgård, H. K. (2013) The definition of culture in culture-based urban development strategies: Antagonisms in the construction of a culture-based development discourse. *International Journal of Cultural Policy*, 19(2), 182–200.

Mangset, P., Kangas, A., Skot-Hansen, D., and Vestheim, G. (2008). Editors' introduction: Nordic cultural policy. *International Journal of Cultural Policy*, 14(1), 1–5.

McGuigan, J. (2004) *Rethinking Cultural Policy*. Maidenhead: Open University Press.

Memorandums from the Program Seminar for Innovation Platforms (Memorandums) (2014) Malmö, 22–23 January 2014.

Mercer, C. (2002) *Towards Cultural Citizenship: Tools for Cultural Policy and Development*. Stockholm: Bank of Sweden Tercentenary Foundation.

Nolin, J. (2010) Sustainable information and information science. *Information Research*, 15(2). Available at: www.informationr.net/ir/15-2/paper431.html [Accessed 28 February 2014].

Ståhlbröst, A. (2008) *Forming Future IT: The Living Lab Way of User Involvement*. Luleå: Luleå University of Technology, Department of Business Administration and Social Sciences, Division of Informatics.

Stevenson, D. (2004) "Civic gold" rush: Cultural planning and the politics of the third way. *International Journal of Cultural Policy*, 10(1), 119–131.

Stevenson, D., Rowe, D., and McKay, K. (2010) Convergence in British cultural policy: The social, the cultural and the economic. *Journal of Arts Management, Law and Society*, 40(4), 248–265.

Strom, E. (2003) Cultural policy as development policy: Evidence from the United States. *International Journal of Cultural Policy*, 9(3), 247–263.

Sundin, M. (2007) *Bra läge men dåligt rykte: En jämförande historisk studie av tre stadsdelar i Borås, Eskilstuna och Gävle* [*Good Location but Bad Reputation: A Comparative Historical Study of Three City Districts in Borås, Eskilstuna and Gävle*]. Uppsala: Uppsala Universitet, Sociologiska institutionen.

Throsby, D. (1997) Sustainability and culture: Some theoretical issues. *International Journal of Cultural Policy*, 4(1), 7–19.

VINNOVA (2014) About VINNOVA [webpage]. Available at: www.vinnova.se/en/About-VINNOVA/ [Accessed 28 February 2014].

Williams, R. (1985) *Keywords: A Vocabulary of Culture and Society*. Rev. ed. New York: Oxford University Press.

13 The town is the venue

"Place-making" at the heart of cultural policy

David Stevenson and Rachel Blanche

Introduction

Representative of a wider shift in urban planning and community development, cultural policy in Scotland has increasingly adopted the idea of "place-making" and considered the part that culture might play in this activity. As part of the Scottish Government's implementation of a "Single Outcome Agreement" between national and local government, a commitment to the development of "sustainable places" has been enshrined as one of the fifteen "national outcomes" for the country (SG 2012). These places are understood as locations where people want to live, in part because of the degree to which they encourage "creative activity and social interaction by providing easy access to both cultural amenities and green spaces" (SG 2012). Indeed, in developing its new strategic plan in 2013, Creative Scotland (Scotland's development body for the arts and creative industries) fully embraced this notion and the central role it perceived itself as having in its delivery. It set out a vision of a country where "places and quality of life are transformed through imagination, ambition and an understanding of the potential of creativity" (CS 2014a, 17).

This chapter considers a practical example of what has been celebrated as successful "place-making" through offering a rich case study of one of the winners of Creative Scotland's Creative Place Awards – Huntly in Aberdeenshire. Beginning with an overview of the award, the chapter goes on to describe some of the activities that have been undertaken as part of an ongoing reimagining of the town as a venue. The research is based on an analysis of secondary sources, observational ethnographic approaches, and a series of in-depth interviews.[1] In presenting this case, we question the degree to which *place-making* is an appropriate description of what has been occurring. Reflecting on both the local policy context and the importance of the "vernacular culture" (Gilmore 2013) of place and the "structures of feeling" (Williams 1977) it informs, the chapter argues that the reason the "cultural interventions" in Huntly have been seen as successful is that they have all been closely connected to the lived realities of the populace and rooted in the *culture-as-way-of-life* with which they are familiar. Rather than prescribing pre-legitimated "culture" to a "knowable community" (Williams 1977) in need of transformation and assistance to ensure its "sustainability," cultural activity, both new and existing, has been combined and contrasted to provide a lens through which both the community and the culture of Huntly can be perceived from a new perspective. In doing so, this activity has encouraged both recognition and celebration of the indigenous culture that was already there and which has been integral in creating the "sense of place" that the residents of Huntly have.

Scotland's Creative Place Awards

In 2012, Creative Scotland launched the Creative Place Awards, which were intended to "celebrate and recognize the hard work and imagination that contribute to the rich cultural life of a community, as well as its social and economic well-being" (CS 2013a). David Taylor and Lulu Johnston, who were responsible for the awards at Creative Scotland, explained that the initiative focused on creative activity in areas beyond Scotland's main cities, in particular seeking applications from organizations and places that may have never applied to an arts council for funding or support before. They explained that the notion of focusing arts funding on a geographical place rather than a demographically defined "community" was a new approach and that at first there was uncertainty about how it should work. The intention was to highlight activities taking place below Creative Scotland's normal radar with an award that celebrates the place – the town as a whole and all that was happening within it – rather than defined segments of the community or the activities of a singular organization or artist.

It is arguable that with this award, Creative Scotland also sought to recognize and support "vernacular culture" (Gilmore 2013) rather than to impose "legitimate" cultural activity where those in positions of power deemed it to be lacking, an approach that Stevenson (2013) argues has been the prevailing discourse of cultural participation policies in Scotland. Policy initiatives like the Creative Place Awards, which begin with such a positive assumption about the populace to whom they are directed, are not typical of cultural policy in Europe, which has tended towards a "deficit" model (Jancovich 2011; Miles and Sullivan 2010). Indeed, it was essential that applications for the award demonstrated preexisting roots of an ongoing "high quality and distinctive creative programme of work" that would take place independently of the award being won (CS 2013b). The award is thus notable because of its apparent rejection of an approach in which an injection of "culture" is provided to a "knowable community" (Williams 1977) deemed to require the positive externalities that cultural activity is presumed to offer.[2] Instead, the award is intended "to enhance and promote the [full spectrum of] work" (CS 2013b) that self-identifying communities are already managing to produce, sustain, and engage with in their local area.

Huntly – a case study of a "creative place"

Huntly is a small town in the northeast of Scotland with a population of 4,500 that has little in the way of any substantial bricks and mortar cultural infrastructure and no prospects of being able to invest in the sort of cultural "hardware"

(Landry 2013) that has often been the defining feature of cultural regeneration and sustainability projects in larger urban centres.[3] Arts organizations can often become venue-focused, and sustaining the venue at all costs may inadvertently become the primary objective for those charged with its management, even when its continued existence may be to the detriment of the diversity of culture that exists around it. However, as Claudia Zeiske, Director of Deveron Arts,[4] points out, in Huntly they do not face that problem as they have undertaken a reimagining of the existing urban landscape through which the *town is the venue*. Any part of it can act as studio, gallery, and stage as required for artists of all disciplines invited from around the world to live and work there.[5] Engaging with local people, their communities, and the existing cultural activity of the town through topics of both local and global concern, the ongoing programme of activity that accompanies this reimagining seeks to bring together artistic and social relationships in a global network that extends throughout and beyond the geographic boundaries of Huntly (DA 2013d). In doing so, it is wholly concerned with what Landry (2013) has called the "software aspect" of urban planning, that which is concerned with "how a place feels, its capacity to foster interactions and to develop and harness skill and talent" (6).

Artist residency programmes often have the artist as the starting point and then try to establish some relevance for the target community, more often than not based on demographics. But Deveron Arts often establishes a relevant theme first, an issue that will engage local people, and then selects an appropriate artist on that basis. Furthermore, they do not work with any particular social groups, but, instead, they work on the basis that they want to engage with the whole town. Zeiske stresses that context is the starting point; the work is about discussion, debate, and dialogue. The "cultural intervention" is not about a transaction but about a relationship. According to one interviewee, without the artistic programme of Deveron Arts, the town would not have explored the issues it has done, and with such a broad variety of local groups and individuals.

50/50 approach

The inception of this reimagining lies in Deveron Arts, a contemporary arts organization that has been based in Huntly since 1995 when its three founders established it as an answer to what they originally perceived as a lack of "cultural activity" in the area (Zeiske and Sacramento 2010, 23). The cofounders have been integral in implementing what they describe as the "curatorial methodology" that reimagines the town as the venue, and bringing both UK and international artists to take part in this process. Deveron Arts' collaborative, socially engaged practices span disciplines and sectors in order to explore the relationship between artist and community, and, in doing so, the relationships within the community themselves. The organization has adopted a "50/50 approach" where "the relationships between local/global, artist/community, hospitality/criticality frame [its] work" (DA 2013d). For example, when many within the town were both individually and publicly debating about if, and where, to locate wind turbines in the locality, this was concomitantly explored alongside debates about global climate change in a piece of work by Scottish artists Matthew Dalziel and Louise Scullion (DA 2013d). The creative practice that visiting artists undertake never is judgmental but, rather, is intended to raise and explore issues that have resonance from both a local and a global perspective. The activities consider the place within the community of those who take part in the activities as well as the place of the community within the wider world. In reflecting on the key strengths of Huntly's bid, Creative Scotland acknowledged the extent to which international links had allowed local concerns to find a synergy with those happening elsewhere in the world (CS 2013c).

The Town Collection

A legacy of the various collaborations between visiting artists and residents of the town is "The Town Collection" of artworks displayed in everyday locations around Huntly. Each residency undertaken contributes another artwork or artefact to what is "an expanding living archive" (DA 2013c). These works are not held in one place, such as a gallery or art center, of which there are none, but instead they are housed in local shops, businesses, public buildings, and private residencies throughout the town and within the surrounding area. The over sixty works in the collection are physical reminders of particular relationships and interactions that reflect the diversity, interests, and characters of the local residents as much as they do the agendas and aesthetics of the artists who have worked in the town.

Local identity and "sense of place"

Some places in the UK are often discursively dismissed as cultural "cold spots"; in an article about "crap towns," Gilmore (2013) has argued that most often "cultural strategies for the arts relate to, or ignore, the specificities of places, the situated cultural practices and implicit knowledge of localities, their internal logics, histories and structures" (86). The everyday realities of the inhabitants become obscured by "official knowledge which privileges legitimate forms and institutions and neglects local contexts of participation" (94). The antithesis to this approach is arguably Landry's notion of *creative city making* (see e.g. Landry, in this volume). While Landry's work on the creative city has become interwoven with the "creative class" thesis of Richard Florida (2002) that sees the attraction and retention of those with significant preexisting cultural capital as a route to urban regeneration and economic growth, this focus on securing an influx of "new" creativity overlooks the value that Landry places on the creativity already present in any given location. He stresses that *creative city making* is predicated on "unleashing the creativity of citizens, organizations and the city" and that it "searches out what is distinctive and special about a place" (Landry 2013, 6). As the experience of Huntly suggests, if interventions of this nature are to succeed, then they must be as concerned about the situated, vernacular, and ostensibly quotidian culture of those who live there and for whom any "sense of place" will always be contingent on the lived reality of the day-to-day.

Certainly, Huntly has a strong, historical local cultural heritage, and interviews highlighted the extent to which people were proud of their own skills and interests, be that dancing, knitting, cooking, or otherwise. A number of these manifestations of what Gilmore (2013) would understand as "vernacular culture" were mentioned in interviews, including bagpiping, dancing, and amateur dramatics in addition to annual events such as the summer ceilidhs, gala days, Highland Games, and regular entertainment in the town square. Indeed, glancing at the notice boards in the windows of local shops, one can identify a range of groups and activities including tea dances, photographic exhibitions, and craft fairs.

The Creative Place Award was very much about recognizing this "sense of place," and many of the artists who came to work in Huntly directly explored the identity of the town through their work; as one interviewee stated, "[W]hat Deveron Arts has done is to bring a contemporary slant to an already rich cultural heritage." One such example would be *Pride of Place*, delivered in July–August 2013 in collaboration with the Caravan Gallery (DA 2013b). The Caravan Gallery was placed in the town square and exhibited the artists' bespoke Huntly collection. Throughout their time there, the artists engaged with passersby, collecting local stories and compiling a map of the town that comprised memories and stories linked to physical locations, thus documenting cultural knowledge and memory in a manner that acknowledged the sense and specificities of place with which it is inextricably linked.

Perhaps the most explicit manifestation of exploring and explicating this individual and collective "sense of place" was in the work by South African artist Jacques Coetzer, which was concerned, quite explicitly, with local people engaging in a process of articulating the town's identity in order to establish a "brand image." Informed by a community consultation and commissioned by the Aberdeenshire Towns Partnership in collaboration with Aberdeenshire Council and Deveron Arts, the two primary outputs from the project were a new slogan for the town – *Room to Roam* – and an accompanying logo.

The town website states that the *Room to Roam* concept represents "[a] motto, a place, a point of view"; it "describes an open, unrestricted landscape and recognizes our individual life journeys towards a shared destination" (Huntly 2014). The logo – found prominently across the town through branded merchandise for sale – is derived from connecting "nodes" representing the characteristics of the town as identified by those who took part in the artistic work: its diversity, its relative geography, its location at the convergence of two rivers, and an existing local tradition of antler imagery. This new "brand identity" has since inspired the annual Room to Roam Festival, which takes this graphical representation of local identity as a starting point for an exploration of landscape and place through a variety of events, walking activities, and celebrations.

"Disturbing the earth"

Despite the innovative and engaging nature of these projects, it is important to note that interviewees also felt that much of the cultural activity that had been taking place in Huntly was not viewed as *culture* by those who took part in it; instead, the activities were simply perceived as "things you do because you enjoy them." The activities of Deveron Arts and, more specifically, the involvement of the artists the organization worked with thus offered a new lens through which to consider what was already there. Concerning the extent to which Deveron Arts could be argued to be "curating" culture in Huntly, and with its almost subcontractor relationship with the local authority at risk of becoming a single filter through which culture is both understood and supported, representatives of Aberdeenshire Council prefer to use the metaphor of "disturbing the earth" to describe what they have seen taking place. It is clear that despite the publically stated motivation for establishing Deveron Arts by its founders, they perceive both it and Zeiske more as bridge builders than as gatekeepers. Certainly, it is evident that all of the visiting artists start from the "structures of feeling" (Williams 1977) that exist rather than seeking to identify that which is lacking. They point to the *Ceidlidhamba* as an example. While a local ceilidh may happen, it can easily become "taken for granted." However, bring in a Brazilian samba band, and it becomes instantly internationalized and offers a new perspective on a local tradition. In doing so, they feel it shines a spotlight on what might be seen as everyday, thus helping to both legitimate and celebrate the indigenous and vernacular cultural activity upon which the particular project has been built.

In undertaking this research, it became clear that it was not only the work of Deveron Arts that had secured the Creative Place Award. In its evaluation of the bid, Creative Scotland highlighted the extent to which Deveron Arts works alongside and complements the wide variety of organizations and individuals in the area – cultural or otherwise – that, as a network, were seen to be providing a diverse variety of year-round opportunities for local residents to participate in, or partake of, contemporary, traditional, and classical work. Indeed, a number of the informants were keen to point out that many of the indigenous cultural activities (bagpiping, highland dance, traditional crafting, etc.) are stronger than ever, and the town is experiencing a surge of interest in people wanting to take part. While it may be true that this surge is in some part thanks to Deveron Arts, it is apparent that if the organization was to cease its work, these activities would still continue, although admittedly less visible to policy makers and the broader "cultural sector." While recognizing the central role of Deveron Arts, interviewees felt that the award endorsed and celebrated *everything* that went on in the town, not just that which happened under the rubric of *The Town is the Venue*.

This respect for the indigenous and vernacular can be seen in the manner in which the £100,000 Creative Place Award money was spent. In addition to developing the *Walks of Life* project that explored the idea of "slow walking" in one of the town's most inescapable contextual features, the surrounding countryside, it was also used to resource a cultural fund for creative groups in and around the town to draw upon (DA 2013a, 13). This supported a diverse range of activities, organizations, and individuals, including many of those highlighted above. This plan to distribute smaller funds to a broad spectrum of organizations and individuals in the area, so as to support an eclectic gamut of cultural activities, was a further area of strength recognized by the Creative Scotland team (CS 2013c).

It should also be noted that what was observed was not a one-way transaction in which only the town and its cultural life was altered by the relationship with the visiting artists. While the town may be the venue, it is not a white cube gallery, and Lorraine Grant and Kirsty Duncan of Aberdeenshire Council have seen the extent to which the town has permeated the work and shaped its legacy. Some of the projects have become regular features of the town's cultural programme, albeit with adaptations and alterations as the inhabitants of the town have taken more direct ownership of their delivery. For example, while for many the *Huntly Hairst* festival initially felt like an "arts event," now it happens because people continue to get enjoyment out of it. This acceptance of the diversity of ways in which the town's inhabitants will engage with the "cultural output" of projects is arguably central to the long-term success of Deveron Arts' work; people can be involved in the activities whether they perceive them as art or simply as an enjoyable activity. Residents are free to participate for whatever reason they want to, and there is no *expectation* for them to be transformed, enriched, or improved.

Although differing activities have generated varying levels of interest, enthusiasm, and understanding, all of them have become part of the collective memory of the town. All the local people spoken to during the fieldwork (either formally or informally) were aware of one event or another, and there were varied opinions regarding their relative worth and merits. Indeed, this diversity of opinion and response is welcomed as evidence of the degree to which the inhabitants of the town have engaged with the project. In order for the work to be seen as successful, there is no necessity for the town to project the sort of fabricated "cozy togetherness" that Landry argues obscures the inherent conflicts and differences that will exist in any community. The curatorial methodology appears to address his assertion that these sort of "place-making" activities should seek to "acknowledge the conflicts [and try to] negotiate a way through them" based on a desire to live with diversity and difference (Landry 2013, 5).

Putnam has argued that in any attempt to increase civic participation, "actions by individuals are not [in themselves] sufficient . . . they are necessary" (Putnam 2000, 403). However, in the case of Huntly, this might be adapted to argue that actions by individuals are necessary in order to fully leverage the potential of the civic participation that is already taking place. Despite the clear evidence that cultural activity and networks did exist prior to the arrival of Deveron Arts, it must be acknowledged that this organization, and in particular its director, has been integral in forging a link between these networks of vernacular or indigenous culture and those networks that constitute what is discursively understood as the "cultural sector" (Gray 2000; Stevenson, 2014). The evaluation of Huntly's bid makes clear that Zeiske was seen to provide a "strong sense of creative leadership in the town" (CS 2013c), and the team at Creative Scotland made it clear that gaining the award hinges on a community having this sort of leadership: someone who can lead the application, forge or identify the necessary partnerships, and translate all of the cultural activity that is happening in a manner that relates it to the objectives that the award was established to achieve.

Local policy context

Despite administering an area close to 2,500 square miles, Aberdeenshire Council is the only local authority in Scotland to be headquartered outside the borders of its own administrative area. This is because, despite its size and population of around a quarter of a million people, there is no large urban centre with more than eighteen thousand residents. This geographical factor has impacted much of the Council's work, and that of the department of Leisure, Culture and Tourism is no exception. Speaking with representatives from the department, it becomes apparent that due to the absence of a distinct "hub location" at which to target funding, investments, and resources, the Council has historically managed its cultural provision at what is perceived as a very "grassroots" level. They see its job as facilitating "embryonic community activity," which has resulted in a partnership approach becoming fundamental to the development and implementation of the region's cultural policy. As such, they regard Huntly not as exceptional but rather as a particularly good example of this approach. According to Grant and Duncan, for the local authority, the appeal of what Deveron Arts has been undertaking is that everything is about the town, its residents, and their lives, meaning that every project undertaken is "relevant, embedded, and sustainable."

Although Aberdeenshire Council, like central and local governments across Europe, seeks a range of outcomes from its spending on culture – not least, economic development, health and wellbeing, education, and tourism – the local authority appears to value independent or "grassroots" cultural activity in preference to trying to provide or curate it itself. As such, Duncan and Grant believe the role of the Council is not about providing *culture* but rather about providing *support*. While that support will vary depending on the location and budget available, it might feasibly include access to buildings, sourcing equipment, supporting promotion, or providing encouragement and expertise. However, what is foremost in the decision-making process is a consideration of what appears to be the most efficient way to use limited resources to help extant cultural activity to continue or nascent practice to take root. Because each community is different, with a very strong local identity, the local authority has to work differently with different communities in order to best meet their cultural needs. As such, while Duncan and Grant are clear that the Council does have a duty to ensure there is "cultural expertise" in the area, it is secondary to community values, satisfaction, and demand. If the residents are not happy with the quality of local cultural activity, then the Council is not fulfilling its role as effectively as it might.

Partly in acknowledgement of the Scottish Government's desire to see a greater focus on "sustainable places" and partly as a response to continuing budget constraints, Aberdeenshire Council has opted to build a new cultural strategy that has "place-making" at its heart and that has been informed by the success of working with the community in Huntly, and with Deveron Arts in particular.

The strategy, entitled *Be Part of the Picture*, is focused on identifying, reinvigorating, and refreshing the identity of the region, and it will see funds being made available of up to £5,000 to support local organizations and individuals to undertake creative and cultural activity in their area (AC 2013). Furthermore, Deveron Arts has been active in sharing its approach with other towns across Scotland and around the world. Indeed, in 2014 one of the Creative Place Award winners was Helmsdale, a village on the east coast of Sutherland. Its application was led by the organization Timespan, whose director, Anna Vermehren, was previously project manager at Deveron Arts and who has arguably adopted a very similar approach since moving further north. On receiving the award, she stated that this work aimed to make "visible the often unseen and unrecognized creativity of Helmsdale's inhabitants while supporting the development of significant new work across multiple art forms by established creative practitioners" (CS 2014b).

Conclusions

So what can be learned from a small community in what many might dismiss as the rural wilds of Scotland? This chapter suggests that in the success of Helmsdale lies the promise: that what has been done in Huntly might be replicable elsewhere; that this particular approach, taken in a small town in the north of Scotland, may have as much application in a village in Spain as it would in a large town in Portugal or a specific neighbourhood in Berlin. Indeed, in Zeiske and Sacramento's (2010) book *ARTocracy*, this very question is posed but never explicitly answered. After conducting this case study, the authors would not only answer this question in the affirmative but argue that it is necessary, not only because of the close interaction it demands with the specificities of extant context and culture, but because this interaction ensures that those activities supported with public money are as concerned with the cultural politics of place as they are with the politics of culture.

Arguably, labelling the "town as the venue" perhaps underplays both the complexity of the place and its centrality to supporting cultural activity in the manner described above; as one community member stated, "[I]t's more than the venue, it's the culture; the audience; the context." All of the projects, products, and practices undertaken by Deveron Arts become interwoven traces within the lived experience of the town's residents that combine to provide the "structures of feeling" (Williams 1977) that arguably transubstantiate the physical and sociocultural context in which a group of individuals live in a perceived sense of "place" that may well be shared in whole, or in part, by the particular communities that live in the locale. In turn, the culture that the community "produces" is both representative of and reactive to the idea of place they have constructed at that particular time.

Seen in these terms, the idea of sustainable *place-making* becomes both problematic and doctrinaire. It appears to assume some sort of *tabula rasa* upon which a *place* will be created only through the targeted interventions of those who seek to "make" it, somewhere that is judged to be "good" by those who are not "of that place" and that once "made" is expected to remain that way evermore. The discourse to which this relates is the same as that which tends to refer to the *sustainable city* in the abstract, a geographical everyman that needs to be manufactured, protected, and maintained at all costs. Yet, *sustainability* is an equally troublesome notion when dealing with the lived reality of place – be it a city, town, or village. What exactly is to be sustained? Who should decide? And how does one identify the difference between decline and change if Doreen Massey's configuration of place is accepted as "never finished, open-ended, interconnected trajectories which make and re-make place, [the evidence for which] is located in the 'throwntogetherness' of place – the constellation of elements of natural and social origin which come together in the here and now" (Massey, cited in Anderson 2008, 232)?

As Grant and Duncan made clear, the work in Huntly has required a foundation of trust to be established within the community, of which Deveron Arts and the team that works there are very much a part. While artists may come from across the globe, their interventions are facilitated by people whom residents might regularly see in the neighbourhood shop. Local people are involved at every stage of a new intervention, and both this process and the methodology that informs it ensure that the objects, performance, actions, or events produced are reflective of the broad spectrum of local interests and issues at that time, and they acknowledge the diversity of the public sphere in which it takes place. As such, anyone wishing to replicate what has happened in Huntly should approach it not with the mindset that they are asking communities to participate in "culture" but instead with the mindset that they are asking what the culture of the community is and how they and their visiting "culture" might participate in it for whatever duration they are present. Therefore, policy makers may need to stop asking how *culture* might be used to create a *sustainable community* and, instead, start looking at how communities have sustained their culture despite being overlooked by many of the institutions and bodies that are charged with supporting it.

Notes

1 In-depth interviews were conducted with Claudia Zeiske, Director of Deveron Arts; Fiona Massey, Chair of the Huntly and District Tourism Action Group; Lorraine Grant, Cultural Services Manager Aberdeenshire Council; Kirsty Duncan, Acting Cultural Services Manager at Aberdeenshire Council; David Taylor, Portfolio Manager at Creative Scotland; Lulu Johnston, Development Officer at Creative Scotland; and Donald Boyd, Director of Huntly Development Trust.

2 See Belfiore and Bennett (2010) for an overview of the literature regarding the social impact of the arts and a discussion concerning the validity of these assumptions.

3 A local example of this is the £48.5 million investment in establishing a Scottish outpost of the Victoria and Albert Museum (Carrell 2013).

4 Deveron Arts was the lead organization in Huntly's application for the Creative Place Award.

5 It should be noted that Deveron Arts makes use of small offices (part of the local library) that are provided by the local authority.

References

Aberdeenshire Council (AC) (2013) Innovative placemaking scheme to revitalise Aberdeenshire image unveiled. News release, 28 May 2013. Available at: www.aberdeenshire.gov.uk/news/release.asp?newsID=3107 [Accessed 24 June 2013].

Anderson, B. (2008) Doreen Massey "for space" [2005]. In P. Hubbard, G. Valentine, and R. Kitchin (eds.), *Key Texts in Human Geography* (pp. 225–235). London: Sage.

Belfiore, E., and Bennett, O. (2010) *The Social Impact of the Arts*. Basingstoke: Palgrave Macmillan.

Carrell, S. (2013, January 8) Dundee V&A will not open until 2015 at the earliest. *Guardian Online*. Available at: www.theguardian.com/artanddesign/2013/jan/08/v-and-a-scottish-project-delayed [Accessed 3 February 2014].

Creative Scotland (CS) (2013a) *Creative Place Awards*. Available at: www.creativescotland.com/explore/projects/creative-place-awards [Accessed 1 August 2013].

Creative Scotland (CS) (2013b) *Creative Place Awards Guidelines 2013/14*. London: Creative Scotland.

Creative Scotland (CS) (2013c) *Key Strengths of Huntly's Applications to the Creative Place Awards*. Documentation provided to the authors by Creative Scotland on 16 August 2013.

Creative Scotland (CS) (2014a) *Unlocking Potential, Embracing Ambition: A Shared Plan for the Arts, Screen and Creative Industries 2014–2024*. Edinburgh: Creative Scotland.

Creative Scotland (CS) (2014b) Winners of the 2014 Creative Place Awards announced (29/01/2014). News release. Available at: http://archive.creativescotland.com/news/winners-of-the-2014-creative-place-awards-announced-29012014 [Accessed 19 November 2014].

Deveron Arts (DA) (2013a) *Deveron Arts: The Town Is the Venue, Annual Report 2012–13*. Huntly: Deveron Arts.

Deveron Arts (DA) (2013b) *The Caravan Gallery: Pride of Place Project*. Available at: www.deveron-arts.com/caravan-gallery/ [Accessed 20 August 2013].

Deveron Arts (DA) (2013c) *The Town Collection*. Available at: www.deveron-arts.com/town-collection/ [Accessed 4 February 2014].

Deveron Arts (DA) (2013d) *The Town Is the Venue*. Available at: www.deveron-arts.com/about/the-town-is-the-venue/ [Accessed 4 February 2014].

Florida, R. (2002) *The Rise of the Creative Class: And How It's Transforming Work, Leisure, Community and Everyday Life*. New York: Basic Books.

Gilmore, A. (2013) Cold spots, crap towns and cultural deserts: The role of place and geography in cultural participation and creative place-making. *Cultural Trends*, 22(2), 86–96.

Gray, C. (2000) *The Politics of the Arts in Britain*. Basingstoke: MacMillan.

Huntly (2014) *Room to Roam: A New Identity for Huntly: Feedback* [webpage]. Available at: www.huntly.net/roomtoroam/feedback.php [Accessed 4 February 2014].

Jancovich, L. (2011). Great art for everyone? Engagement and participation policy in the arts. *Cultural Trends*, 20(3–4), 271–279.

Landry, C. (2013) Culture: Key to sustainable development. Keynote address at the UNESCO International Congress "Culture: Key to Sustainable Development," 15–17 May 2014, Hangzhou, China.

Miles, A., and Sullivan, A. (2010) *Understanding the Relationship Between Taste and Value in Culture and Sport*. Available at: www.culture.gov.uk/publications/7542.aspx [Accessed 25 February 2014].

Putnam, R. D. (2000) *Bowling Alone: The Collapse and Revival of American Community*. New York: Simon and Schuster.
Scottish Government (SG) (2012) *National Outcomes*. Available at: www.scotland.gov.uk/About/Performance/scotPerforms/outcomes/susplaces [Accessed 25 February 2014].
Stevenson, D. (2013) What's the problem again? The problematisation of cultural participation in Scottish cultural policy. *Cultural Trends*, 22(2), 77–85.
Stevenson, D. (2014) Tartan and tantrums: Critical reflections on the Creative Scotland 'stooshie.' *Cultural Trends*, 23(3), 178–187.
Williams, R. (1977) *Marxism and Literature*. Oxford: Oxford University Press.
Zeiske, C., and Sacramento, N. (2010) *ARTocracy: Art, Informal Space, and Social Consequence*. Berlin: Jovis.

Section IV

Making the city resilient

Building communities through artivism

14 Mobilizing urban neighbourhoods

Artivism, identity, and cultural sustainability[1]

Milena Dragićević Šešić, Aleksandar Brkić, and Julija Matejić

Introduction: "small" cities in a "small" country

The time period of the late twentieth and early twenty-first centuries has brought significant changes and tough challenges to postwar Yugoslav countries (i.e. Wachtel 1998; Dragićević Šešić and Dragojević 2006; Švob-Đokić 2011; Brkić 2014). Forced and willing migrations, intensified ethnic distances, destroyed industries, and the combined consequences of globalisation, the economic crisis, and the process of European integration have changed not only the ingrained way(s) of life but also the appearance and perception of life in urban environments in former Yugoslav countries.

Within transitional[2] Serbia, numerous city centres and neighbourhoods have lost their identities due to a growing sense of alienation and a diminishing sense of shared place and locality. This has occurred as a result of three main factors: the collapse of urban industrialisation and the idea of shared/common spaces from the socialist period, dissatisfaction with the current urban environment, and a feeling of disengagement from society. Consumerist culture's shopping malls and other popular *non-places* (Augé 1997) have replaced the public spaces traditionally used for communication and gathering (i.e. *kafana*, *korzo*). Due to this process of *decontextualisation* (Dragićević Šešić 2011b), neighbourhoods are disappearing – now less defined as places, and more as temporary living spaces.

In recent decades, cultural researchers have focused mainly on metropolitan cities (e.g. Bianchini 1993; Johnson 2009; Patel 2013). Furthermore, researchers have tended to overlook and neglect *artivism* as an increasingly present phenomenon of social activism through the arts. *Artivism* is a neologism, a hybrid notion consisting of two words: art and activism. In his seminal essay, "Artivism," Aldo Milohnić (2005) explains that *artivism* is a kind of *interventionism* which uses "cultural-manifestation techniques in order to become constituted in the field of the political" (18). Furthermore, "the transversality of these practices and their hybrid nature enable quick passages from the predominantly artistic into the predominantly political sphere and back" (24). Writing about the *artivist*, Milohnić says that "an activist is an artist as much as is inevitable, no more and no less; the artisanship is a side effect of a political act" (20).

In light of this, the chapter considers what is left of the *once upon a time* diverse identities of small and medium-sized cities (SMCs) in Serbia, today converted to similarly shaped and underdeveloped postindustrial cities and their neighbourhoods, and it observes the potential impact of artivism initiatives in reinvigorating and contributing to the life and sustainability of these urban units. The paper poses two essential questions: to what extent can artivistic practices in public spaces be mobilising? Can artivism initiate a rethinking of neighbourhood identity and lead to social changes and meaningful urban cultural policies? This research is not about image-making and branding – it is about creating sustainable communities through bottom-up artivism and uses of civic imagination.

Rebuilding neighbourhood identity

Neighbourhood is one of the most important concepts for understanding a local community, connected through a common space as its base of relations, with proximity as one of the main characteristics of a neighbour (Bulmer 1986) and the potential for social reproduction (Appadurai 1996). Among the results of the modernization of European societies were changes in the perception and definition of neighbourhoods, where proximity is more of a notion from past times than a constituent characteristic of a neighbour in a contemporary local community, where people do not have enough direct contact with their neighbours or feel that they are in control of the place (Power 2007). As Eric Hobsbawm has said, "[N]ever was the word 'community' used more indiscriminately and emptily than in the decades when communities in the sociological sense became hard to find in real life" (Hobsbawm 1995, 428).

Furthermore, the concept of *neighbourhood* is in direct confrontation with the concept of *consumer society* since it calls for a strong connection with one place and one group of people and for immobility to a certain degree, which degrades contemporary Western societies' focus on movement and superficial human relations that allow easy physical, social, industrial, and emotional mobility. This tension is even more accentuated in the transition to more consumer-based societies, affected by both processes of globalisation and the current neoliberal politico-economic trends of governance in Serbia.

Inspired by Giuliana Bruno's (2007) making of *Atlas of Emotion* by connecting sight and site, motion and emotion, architecture and film, the case study we use in this chapter aims to connect cultural activist performative projects with urban territories – *des ilots urbains*, spaces with specific character within a city. It is an attempt to sustain a project that presents an interplay of urbanism, architecture, performance studies, culture of memory, and cultural management in order to prove that the real sustainability of particular city neighbourhoods is created or achieved through their cultural practices, which rely on the engagement of multiple senses and the spirit of the place. This involves a complex intervening of individual and collective action: from *sightseeing* (touristic consumerism) to *site seeing* (exploration of our own city's meanings), the latter defined as a relation "established between places and events that forms and transforms the narrative of a city: the city itself becomes imagined as narrative as sites are transformed by the sequence of movements of its traveller-dwellers" (Bruno 2007, 66).

We also highlight the importance of the relation between a strong local identity and culture-centred sustainability, which has "positive impacts in creating the preconditions for establishing civic pride, community spirit, and the necessary caring for the urban environment" (Ratna and Piracha 2006, 36). Moreover, as Dominique Moïsi (2009) has foreseen and pointed out, we agree that "the mapping of emotions will become as legitimate and compulsory an exercise as the mapping of geographical realities" (158). Knowing that the predominant emotion in Serbian society (particularly in SMCs) is a *culture of humiliation* (Moïsi 2009), we argue that it can be overcome only through bottom-up civic activism as a precondition for establishing civic pride and civic urbanity (see also Landry, in this volume).

Youth in Serbia: making something out of nothing

Young people in Serbia's SMCs share many commonalities with their generation's global struggle with confusion, uncertainty, apathy, lack of interest, information saturation, unemployment, and poor cultural offer (in terms of both quantity and quality) (Azanjac, Bradić, Krivokapić, and Stojić 2012). What they experience in their everyday lives is a phenomenon already described as "emptiness or claustrophobia, rather than neighbourliness" (Miles 2006, 136).

We may define such a *social vacuum* or *void-ness* as the perpetual presence of the absence and the absence of the presence. As Bergson (1983) argued, there is no absolute void in nature because "we can perceive only the *presence* of one thing or of another, never the *absence* of anything. There is absence only for a being capable of remembering and expecting" (281). Paradoxically, there is more, not less, in the idea of nonexistent neighbourliness. By leaving *the void of itself*, such absence is full of nostalgia for the better and desired yet suppressed by the current reality.

In this light, our research is based on the case study selected to represent the paradigm of youth *capable of remembering and expecting* – the campaign "Openly about Public Spaces" (hereinafter, the Campaign). Consisting of different culture-led initiatives that encourage self-organizing and mobilisation of culture aimed at urban regeneration, the Campaign has been implemented by the civic association Civic Initiatives, since 2009, together with youth offices (public structures) in several SMCs throughout Serbia (Kikinda, Kragujevac, Novi Sad, Niš, Kruševac, Smederevo, Užice, etc.).[3]

The Campaign started by launching an official request to all city governments in Serbia, referring to the *Law on Free Access to Information of Public Importance* and asking for the list of unused public spaces. Given that young people are lacking space for their own activities and creative expression, they wanted to see how to either obtain or occupy these abandoned spaces and make something out of them. This action is based on civic self-organized mapping and sometimes illegal occupation of unused and often devastated public spaces (*squatting*), usually followed by negotiations with local authorities in order to reconstruct places, revitalise neighbourhoods, and make a long-term contribution to the sustainable development of the local community.

Two distinctively different approaches can be seen in their fight for the right to use abandoned public spaces. The first approach focuses on organizing numerous cultural events, raising awareness, lobbying, and negotiating with local authorities. For example, in Kikinda,[4] a group of high school pupils launched an initiative to create the youth centre Pantry (Špajz) – a music room, reading club, and place for different creative activities (workshops, exhibitions, etc.). In 2012, after two years of negotiations and a variety of activities, the administration of the city of Kikinda provided the youth centre Pantry with a space for the next twenty years. This space was a public bathroom initially, with neither water nor electricity, and thus Pantry has to be renovated to meet the needs of the youth centre, and, therefore, at the moment of this article's writing, it is not yet in use. However, young activists continue to fundraise, expanding the network of people supporting them and organizing different cultural events on the streets of Kikinda with the intention of raising awareness of the importance of such initiatives and places.

Similarly, the Planet of the Innovative Youngsters (Planeta inovativaca) is an informal group of young people who gathered in 2010 with a desire to create a youth centre and contribute to the city they live in, Smederevo.[5] They organized numerous street-based cultural events and performances, and, in less than six months, more than fifteen hundred people signed their petition for the opening of a youth centre. Although there are many unused spaces in Smederevo, the local government has not shown any interest in supporting this initiative, even though the adopted *Local Plan of Action for Youth of Smederevo 2010–2014* emphasises the lack of a cultural centre for youth in the city.

In contrast, the second, completely different, approach to raising awareness and obtaining unused spaces can be observed in civil disobedience and subversive practices defined as *squatting*. These examples of unauthorised occupation of abandoned public spaces appeared as a reaction to the nonexistence of a space for citizens to engage in independent, educational, and cultural activities important for the community, even though there are empty and abandoned public facilities, military buildings, and industrial complexes.

Figure 14.1 The exterior/interior of the abandoned military barracks Dr. Archibald Reiss, in Novi Sad, Serbia

Source: Orfeas Skutelis.

Figure 14.2 The activist occupation of the abandoned military barracks Dr. Archibald Reiss, in Novi Sad, Serbia

Source: Danijel Šivinjski.

Figure 14.3 The joint military and police action of evicting the activists from the abandoned military barracks Dr. Archibald Reiss, in Novi Sad, Serbia

Source: Danijel Šivinjski.

After three months of preparation, sending requests to relevant but indifferent authorities and representatives of the army, as well as transparently announcing plans and activities on their website, the activists of the initiative for the Independent Social Centre in Novi Sad[6] entered the abandoned military barracks Dr. Archibald Reiss on Army Day 2011. During the three weeks of the centre's existence in the barracks, the facilities were cleaned and numerous programmes were organized (all free of charge and open to the public), including eighteen educational, creative, and humanitarian workshops; twelve music programmes; nine panel discussions; six film and video programmes; three exhibitions; five literary programmes; and five performances. However, representatives of the army and police then evicted the activists from the military barracks and prohibited further access to the abandoned space. To this day, the space remains empty, and the centre is temporarily situated at another location. Furthermore, negotiations with city authorities are at a standstill following political changes in the city administration of Novi Sad.

Similarly, without its own premises and through various actions and street art events, the independent art production organisation Undercity (Undergrad)[7] in Užice[8] emphasised the importance of using abandoned public spaces on many occasions. In 2008, they occupied a former rifle range in the centre of the city (abandoned for more than fifteen years) by painting the empty walls with *post-graffiti*[9] (Reinecke 2007) and placing a mini skate ramp inside the building. Although the police cleared the space after only ten days, the Undercity activists continue to negotiate with local authorities with the aim of obtaining their own space as well as drawing people's attention to the large number of abandoned and unused public spaces in their city.

As acts of social engagement and protest against the public reality and civic lethargy, these and many other examples throughout Serbia have shown the increased youth demand for neighbourliness. The Campaign appears to embody the generation's dissatisfaction with the social environment and articulates an awakening plea for everyone (else) to participate and get involved – to confront social issues and deficiencies through artivistic projects and, moreover, to find, *conquer, unlock*, and use abandoned public facilities for their cultural expression. In other words, the Campaign's artivism "reappeared out of necessity to fight pressures of the market, governmental directive policies, and indolence and incapacities of the public institutional system in culture" (Dragićević Šešić 2011a, 83).

Nevertheless, a lot of initiatives in Serbian cities depend on local elections, political changes, and affiliation to political parties. Local authorities are largely unapproachable and very slow to make decisions. With few exceptions, they tend to remain deaf to such initiatives. Incapable of seeing the numerous benefits in opening public spaces to young people and delegating certain rights to civic organizations, they prefer to keep the abandoned public spaces empty (Dimitrijević Mišković 2012).

Public art in public space as a public good

Although traditionally public space is defined as a place accessible to everyone at all times (such as town squares, streets, parks, etc.), the military buildings mentioned above and similar facilities inaccessible to the public can be perceived as *quasi-public spaces* due to entry restrictions (Knox and Worpole 2007). Needless to say, a city is not just a set of buildings – public or private spaces. As discussed by Lefebvre (1991), "(social) space is a (social) product" (26), a complex social construction. Therefore, every society produces its own unique space(s). Cities, neighbourhoods, and public spaces are not given *per* se – they are socially and culturally constructed because people make places (Mean and Tims 2005).

Rooted also in Lefebvre's *right to the city* concept, the Campaign promotes the importance of neighbourhood place-making, which might be conceived as a *right to the neighbourhood* concept. The Campaign belongs to the public art subgroup of *new genre public art* – artistic expressions are not only situated outside conventional art spaces (galleries, theatres, etc.) but specifically committed to activate citizens' engagement with the environment and participation in the public sphere (Tornaghi 2008, 4). Based on civic engagement, active citizenship, and conflict as creative tools, these initiatives are an attempt to restore the city's significance to its inhabitants. However, the essence is to debate not what public space or public art is but rather what makes a (public) space *public* and how the public art in a public space influences the *public*.

Through artivistic initiatives in Kikinda, Smederevo, Novi Sad, and Užice, in settings that have lost their urban narratives during the transition period, inhabitants are trying to (re)claim control over the process of the production of (public) spaces and their joint use. By attributing particular, new meanings to abandoned spaces and breathing life into them, these youth initiatives attempt to convert *non-places* into *places*.

While the idea of *public* (public good, public interest, etc.) may be shrinking in contemporary society (Cruz 2012), it is increasingly necessary to reformulate and reinterpret its present-day meaning by recognising *civic imagination* in itself as a public good. It is in the public interest to support active citizenship and to stimulate direct cultural participation, which is the cornerstone of the Culture 3.0 paradigm we are experiencing today, "a situation in which individuals do not limit themselves to absorb passively the cultural stimuli, but are motivated to put their skills at work: creating their own programmes and projects, not only responding to Government policies and practices" (Sacco 2011, 5). Thus, we believe that civic imagination is crucial for practicing the *right to the neighbourhood* concept.

Keeping in mind the relationship between public art and the broader social and political environment, the Campaign can be observed in accordance with Kester's (2005) definition of *dialogical* aesthetic, which reflects the notions of *community-specific* (cf. Miles), *relational* (cf. Nicolas Bourriaud), and *conversational* (cf. Homi Bhabha) art practices. Since they are based on dialogue, process, participation, communication, and exchange, we may argue that such aesthetic experiences and artivistic practices can generate remarkably influential changes in the consciousness of their participants who "share insights, and forge a provisional sense of collectivity" (Kester 2005, 84).

Thus, what artivistic initiatives as part of the Campaign might bring (back) to the community is a sense of belonging to a place and connectedness in relation to both other inhabitants and the shared physical environment – that is, neighbourliness. Through processes of coming together with other people as well as the unique spatial and temporal synthesising of self, others, space, and time (*hic et nunc*), the initiatives contribute to the long-awaited process of place-making and meaning-making of abandoned public spaces and silenced neighbourhoods. By stimulating and mobilising citizens to actively participate in urban regeneration and development, as well as encouraging communication, social interaction, and negotiations with local authorities, the Campaign is a potential platform for rethinking and re-creating neighbourhood identities. Therefore, the initiative may have a vital role in the social life of local communities and, moreover, their sustainable development.

A bottom-up approach towards sustainable cultural development

In times of crisis of democracy or, more precisely, of democratic capitalism (Streeck 2011), when power is concentrated in small circles of various nomenclatures, the idea of going back to the *bottom* – where the *real people* are, as presented through the examples from the Campaign – seems like a feasible option but often like a naive one too. It can even seem like an attempt to go back to mythical beginnings, from where everything starts, where all people and events are more *true*. However, localism is more than a meaningful principle and form of social organization, and today it can mean much more. We can even say that "there is no cosmopolitanism without localism" (Blackshaw 2010, 86) while making a distinction between the cosmopolitanism of the elites and the localism of people (Castells 2004). In the same way that we perceive a large city completely differently when we explore it by foot and by underground public transport, local community can be perceived, or perceive itself, as a physically close and contingent group through space/time but also as a neo-tribe (Maffesoli 1996) perceiving the space/time as pointillist and noncontingent, and experiencing life in episodes (Bauman 2007).

Knowing that cities are a unique form of natural, built, and cultural environment (Haughton and Hunter 2003, 9), we may argue that culture, broadly defined as the way people live together (UNESCO 1998), has always played an important role in bringing them *joie de vivre*. From the model of *the managed city* (after the Second World War), to the *postmodern or postindustrial city* (in the 1970s and 1980s), to the *entrepreneurial city*, we can trace the so-called *cultural turn* (Mercer 2006) in the 1990s and 2000s. Since the publication of *Our Common Future* (the Brundtland Report) (WCED 1987), while the concepts of the *creative city* (Landry and Bianchini 1995) and the *eventful city* (Richards and Palmer 2010) emerged and became widely popularized, they are being increasingly replaced with the sole idea of the *sustainable city*.

Although no single definition of sustainable development has been universally accepted, an increasing number of researchers worldwide have tried to bring together the two key concepts of culture and sustainability in public discourses, policy, and planning – to incorporate the notion of culture as *the fourth pillar of sustainable development*.[10]

As Landry and Bianchini (1995) argue, we cannot create sustainable environments by taking into account only environmental, economic, and social dimensions: "[W]e need to assess how 'feel', ambience, atmosphere and 'soft' infrastructures are created, something which requires different skills from those of planners brought up to think in terms of physical solutions" (13). In this vein, local experiences of sustainable cultural development through artivism have to address people, have to understand how they mix and connect, and must engage and inspire them to participate in this process of more meaningful neighbourhood making.

By making an effort to safeguard abandoned public spaces from further deterioration and take them into consideration in light of the public's cultural memory as well as for the benefit of future generations, the Campaign signifies the power of civic imagination and self-organization, a model of alternative and progressive cultural practices that recognises the importance of culture for local development towards an attractive and sustainable living environment. Whereas numerous postindustrial cities are in decline, mostly due to a *culture of humiliation* (Moïsi 2009) because they lost their geopolitical importance and meaning, both the research literature and this case study have shown that people participating in community activities have stronger ties to them, especially if those activities are of an *alternative*, subversive nature. Furthermore, if attached to them, people are more willing to stay in their neighbourhoods and contribute to their sustained development (Bramley et al. 2010). In this context, we also have to emphasise the importance of relationships among generations in the context of social changes – an issue raised more than six decades ago by Mannheim (1952) in his theory of generations, which is still relevant and applicable to our case study today.

Despite the fact that the long-term goals in the cities of Kikinda, Smederevo, Novi Sad, and Užice are not (yet) achieved, and the fact that citizens are stubbornly being excluded from the policy-making process at the local level, we may note that these initiatives are of great importance for local growth and planning. They raised awareness, introduced critical thinking, made the new paradigms more *real*, created meeting places, rethought neighbourhood identities, and, consequently, initiated negotiations with local authorities. These types of innovative bottom-up initiatives that allow citizens to participate in the planning, social transformation, and revitalisation of abandoned areas are perceived as being the only suitable approach towards sustainable cultural development for the reason that they are not imposed from above. On the other hand, such an approach requires a more sustained agenda for long-term impact. Recalling the processual character of artivistic place-making and meaning-making, we may argue that sustainable cultural development is a demanding and long-lasting process rather than a one-time occurrence – as Franco Bianchini has put it, it is a *difficult art* (Bianchini 1993, 19).

In the context of the global economic crisis, which has proven to be beyond the economy, culture and art(s) seem to lose their focus (Bonet and Donato 2011). Given that most art is political nowadays (Vander Gucht 2006), both culture and artistic engagement are potential catalysts for social sustainability (Colantonio 2011). One of the possible answers to the problems arising from identity-based conflicts in Europe is setting the frame for their resolution in the centres of the local communities (urban and rural, large and small), developing programmes which have a positive effect on the level of intercultural dialogue among different cultures, mainly through initiatives of the civil sector, initiated directly by the communities. The field of culture should be placed in the centre of this frame, with its actors using this space to advance the development of more progressive, holistic, and sustainable communities. Thus, a *culture of humiliation* could be replaced by a *culture of hope*, which, at the present moment, can be developed only within the communities using their own civic imagination and activism. This needs to be done, moreover, not only through much more efficient and necessary self-organization and social engagement but also through support from local authorities, which is essential for long-term sustainability, as shown by the relative lack of success in the presented case study in Serbia.

Notes

1 This research was partially developed within project no. 178012 of the Ministry of Education, Science and Technological Development of the Republic of Serbia, and the Research Fund of LASALLE College of the Arts, Singapore.
2 The word *transition* is used to mark the shift from a society with more social equality and multicultural demographic structure (attempted by a communist/socialist Yugoslavia) to societies with radical social inequalities and more or less mono-ethnic countries (resulting from an attempt to quickly appropriate the values of a liberal capitalist society, such as private property, competitiveness, and entrepreneurialism), in a region that tried out most of the modern political systems at some point – "from empire to revolutionary republic, from multi-national federation to nation state to protectorate" (Močnik 2003).
3 For more information, see Veselinović and Stevanović (2012) and Dimitrijević Mišković (2012).
4 Situated in the autonomous province of Vojvodina, the town of Kikinda has 38,065 inhabitants, while the municipality has 59,453 inhabitants (*National Census* 2011).
5 As the administrative centre of the Podunavlje District, the city of Smederevo has a population of 108,209 (*National Census* 2011).
6 As the second largest city in Serbia, Novi Sad is the administrative seat of Vojvodina, with a population of 231,798 (the administrative area of the city has a population of 341,625) (*National Census* 2011).
7 Undergrad is an NGO developed within the programme Open Highway E-761. For more information on this programme, see www.scp-srb.net/files/OpenHighway.pdf [Accessed 28 February 2014].
8 Situated in western Serbia, Užice has a total population of 78,040 (*National Census* 2011).
9 We use the term *post-graffiti* in order to make a distinction from the traditional *graffiti*, usually perceived as vandalism.
10 Some of the most important contributions are Hawkes (2001), Nadarajah and Yamamoto (2007), Papageorgiou (2007), SALAR (2008), UCLG (2010), and Duxbury and Jeannotte (2010, 2011).

References

Appadurai, A. (1996) *Modernity at Large: Cultural Dimensions of Globalization*. Minneapolis: University of Minnesota Press.

Augé, M. (1997) *Non-places: Introduction to an Anthropology of Supermodernity*. London; New York: Verso.

Azanjac, T., Bradić, D., Krivokapić, Đ., and Stojić, T. (2012) *Youth and Public Policy in Serbia*. New York: International Debate Education Association, Demokratie & Dialog e.V.

Bauman, Z. (2007) *Liquid Times: Living in an Age of Uncertainty*. Cambridge: Polity Press.

Bergson, H. (1983) *Creative Evolution*. Lanham: University Press of America.

Bianchini, F. (1993) Remaking European cities: The role of cultural policies. In F. Bianchini and M. Parkinson (eds.), *Cultural Policy and Urban Regeneration: The West European Experience* (pp. 1–20). Manchester: Manchester University Press.

Blackshaw, T. (2010) *Key Concepts in Community Studies*. London: Sage.

Bonet, L., and Donato, F. (2011) The financial crisis and its impacts on the current models of governance and management of the cultural sector in Europe. *ENCATC Journal of Cultural Management and Policy*, 1(1), 4–12.

Bramley, G., Brown, C., Dempsey, N., Power, S., and Watkins, D. (2010) Social acceptability. In M. Jenks and C. Jones (eds.), *Dimensions of the Sustainable City 2* (pp. 105–128). Future City Series, Vol. 2. London: Springer.

Brkić, A. (2014) *Cultural Policy Frameworks (Re)constructing National and Supranational Identities: The Balkans and the European Union*. Amsterdam: European Cultural Foundation.

Bruno, G. (2007) *Atlas of Emotion: Journeys in Art, Architecture, and Film*. London: Verso.

Bulmer, M. (ed.) (1986) *Neighbours: The Work of Phillip Abrams*. Cambridge: Cambridge University Press.

Castells, M. (2004) *The Power of Identity: The Information Age: Economy, Society and Culture, Vol. II*. Oxford: Blackwell.

Colantonio, A. (2011) Social sustainability: Exploring the linkages between research, policy and practice. In C. C. Jaeger, J. D. Tàbara, and J. Jaeger (eds.), *European Research on Sustainable Development, Volume 1: Transformative Science Approaches for Sustainability* (pp. 35–58). Berlin: Springer.

Cruz, T. (2012) Democratizing urbanization and the search for a new civic imagination. In N. Thompson (ed.), *Living as Form: Socially Engaged Art from 1991–2011* (pp. 56–63). Cambridge: MIT Press.

Delanty, G. (2003) *Community*. London: Routledge.

Dimitrijević Mišković, M. (2012) *Javni prostori i kako ih otključati*. Beograd: Građanske inicijative.

Dragićević Šešić, M. (2011a) Contemporary society, art and public space: Towards the creation of new relational and esthetical territories. In F. Trotte, P. Brunet, and L. Barone (eds.), *Culture & Innovation(s): Europe Seen from the South* (pp. 83–98). Relais Culture Europe, Sostenuto. Available at: www.lemproject.eu/in-focus/news/culture-as-a-factor-for-economic-and-social-innovation [Accessed 8 February 2014].

Dragićević Šešić, M. (2011b) Cultural policies, identities and monument building in Southeastern Europe. In A. Milohnić and N. Švob-Đokić (eds.), *Cultural Identity Politics in the (Post-) Transitional Societies* (pp. 31–46). Zagreb: Institute for International Relations.

Dragićević Šešić, M., and Dragojević, S. (2006) Imagined or real divides. In U. H. Meinhof and A. Triandafilydu (eds.), *Transcultural Europe* (pp. 43–56). London: Palgrave Macmillan.

Duxbury, N., and Jeannotte, M. S. (2010) Culture, sustainability and communities: Exploring the myths. Working paper no. 353, *Oficina do CES*. Available at: www.ces.uc.pt/publicacoes/oficina/index.php?id=2982 [Accessed 8 February 2014].

Duxbury, N., and Jeannotte, M. S. (2011) Introduction: Culture and sustainable communities. *Culture and Local Governance / Culture et gouvernance locale*, 3(1–2), 1–10.

Haughton, G., and Hunter, C. (2003) *Sustainable Cities: Regional Development and Public Policy*. London; New York: Routledge.

Hawkes, J. (2001) *The Fourth Pillar of Sustainability: Culture's Essential Role in Public Planning*. Commissioned by the Cultural Development Network, Victoria. Melbourne: Common Ground.

Hobsbawm, E. (1995) *The Age of Extremes: The Short Twentieth Century 1914–1991*. London: Abacus.

Johnson, C. L. (2009) *Cultural Capitals: Revaluing the Arts, Remaking Urban Spaces*. Surrey: Ashgate.

Kester G. (2005) Conversation pieces: The role of dialogue in socially engaged art. In Z. Kocur and S. Leung (eds.), *Theory in Contemporary Art Since 1985* (pp. 76–88). Oxford: Blackwell. Available at: www.ira.usf.edu/cam/exhibitions/2008_8_torolab/readings/conversation_piecesgkester.pdf [Accessed 30 January 2015].

Knox, K., and Worpole, K. (2007) *The Social Value of Public Spaces*. York: Joseph Rowntree Foundation.

Landry, C., and Bianchini, F. (1995) *The Creative City*. London: DEMOS.

Lefebvre, H. (1991) *The Production of Space*. Oxford: Blackwell.

Maffesoli, M. (1996) *The Time of the Tribes: The Decline of Individualism in a Mass Society*. London: Sage.

Mannheim, K. (1952) [1923] The problem of generations. In P. Kecskemeti (ed.), *Essays on the Sociology of Knowledge* (pp. 276–320). London: Routledge and Kegan Paul.

Mean, M., and Tims, C. (2005) *People Make Places*. London: Demos. Available at: www.demos.co.uk/files/Demos_PMP_Final_02.pdf?1240939425 [Accessed 8 February 2014].

Mercer, C. (2006) Cultural planning for urban development and creative cities. Available at: http://burgosciudad21.org/adftp/Shanghai_cultural_planning_paper.pdf [Accessed 8 February 2014].

Miles, M. (2006) Garden cities and city gardens. In D. Bell and M. Jayne (eds.), *Small Cities: Urban Experience Beyond the Metropolis* (pp. 135–150). Abingdon: Routledge.

Milohnić, A. (2005) Artivism. *Maska*, 20(1–2), 15–25.

Močnik, R. (2003, March 20) Social change in the Balkans. *Eurozine*. Available at: www.eurozine.com/articles/2003-03-20-mocnik-en.html [Accessed 8 February 2014].

Moïsi, D. (2009) *Geopolitics of Emotions: How Cultures of Fear, Humiliation, and Hope Are Reshaping the World*. New York: Doubleday.

Nadarajah, M., and Yamamoto, A. T. (eds.) (2007) *Urban Crisis: Culture and the Sustainability of Cities*. Tokyo: United Nations University Press.

National Census (2011) Belgrade: Statistical Office of the Republic of Serbia. Available at: media.popis2011.stat.rs/2014/eksel/Naselja/1_Stanovnistvo%20prema%20starosti%20i%20polu,%20po%20naseljima.xls [Accessed 8 February 2014].

On the Open Road: Catalogue of the Exhibition (2008). Užice, Beograd: Academica. Available at: www.academica.rs/academica/Na%20otvorenom%20drumu.pdf [Accessed 28 February 2014].

Papageorgiou, F. (2007) *Culture and Sustainable Rural Development*. Euracademy Thematic Guide Series, No. 6. Athens: Euroacademy Association, European Academy for Sustainable Rural Development.

Patel, K. K. (ed.) (2013) *The Cultural Politics of Europe: European Capitals of Culture and European Union Since the 1980s*. London: Routledge.

Power, A. (2007) *City Survivors: Bringing Up Children in Disadvantaged Neighbourhoods*. Bristol: Policy.

Ratna, S. J. B., and Piracha, A. L. (2006) Cultural frameworks. In M. Nadarajah and A. Tomoko Yamamoto (eds.), *Urban Crisis: Culture and the Sustainability of Cities* (pp. 13–50). Tokyo: United Nations University Press.

Reinecke, J. (2007) *Post-graffiti: Between Street, Art and Commerce*. Berkeley: Gingko Press.

Richards, G., and Palmer, R. (2010) *Eventful Cities: Cultural Management and Urban Revitalisation*. Oxford: Elsevier.

Sacco, P. (2011) Culture 3.0: A new perspective for the EU 2014–2020 structural funds programming. European Expert Network on Culture (EENC). Available at: www.eenc.info/wp-content/uploads/2011/07/pl-sacco_culture-3-0_CCIs-Local-and-Regional-Development_final.pdf [Accessed 8 February 2014].

Streeck, W. (2011, September–October) The crises of democratic capitalism. *The New Left Review*, 71, 5–29.

Švob-Đokić, N. (2011) Cultural identities in southeastern Europe – a post-transitional perspective. In A. Milohnić and N. Švob-Đokić (eds.), *Cultural Transitions in Southeastern Europe: Cultural Identity Politics in the (Post-)transitional Societies* (pp. 113–123). Zagreb: Institute for International Relations.

Swedish Association of Local Authorities and Regions (SALAR) (2008) *Culture in the Sustainable Society*. Stockholm: SALAR.

Tornaghi, C. (2008) Questioning the social aims of public art in urban regeneration initiatives: The case of Newcastle upon Tyne and Gateshead (UK). Electronic Working Documents, Global Urban Research Unit, Newcastle University, UK. Available at: www.ncl.ac.uk [Accessed 8 February 2014].

UNESCO (1998) *World Cultural Report: Culture, Creativity and Markets*. Paris: UNESCO.

United Cities and Local Governments (UCLG) (2010). *Culture: Fourth Pillar of Sustainable Development*. Policy Statement. Barcelona: UCLG. Available at: www.uclg.org/sites/default/files/9890675406_(EN)_culture_fourth_pillar_sustainable_development_eng_0.pdf [Accessed 8 February 2014].

United Nations World Commission on Environment and Development (WCED) (1987) *Our Common Future: From One Earth to One World*. Report of the World Commission on Environment and Development. Available at: www.un-documents.net/ocf-ov.htm#1.2 [Accessed 8 February 2014].

Vander Gucht, D. (2006) *What Does It Mean for an Artist to Be Political Today?* ITTACA project booklet. Luxembourg: Fondation Marcel Hicter and Casino Luxembourg.

Veselinović, D., and Stevanović, M. (eds.) (2012) *Otvoreno o javnim prostorima*. Beograd: Udruženje građana za demokratiju i građansko obrazovanje "Građanske inicijative."

Wachtel, A. B. (1998) *Making a Nation, Breaking a Nation: Literature and Cultural Politics in Yugoslavia*. Stanford: Stanford University Press.

15 Land, people, and art

An attempt to renew social identity in the Czech Republic

Miloslav Lapka and Eva Cudlínová

Landscape and land art: an introduction to the concept

In speaking about land art and social identity, we are touching on a place-based identification of community. There are many concepts of land art and of site-specific art and, of course, many concepts about social identity too. The aim of this chapter is not to make a critical survey of these concepts; rather, the intention is to analyze an emerging movement that we have named *land, people, and art* (LPA), which is different from the classical land art created by professional artists, and to make some attempt at synthesis between the two forms. In using the term *land, people, and art,* we are stressing the role of community in the process of renewal of the symbolic landscape. Our aim is to stress the potential function of the land, people, and art for renewal of social identity – that is, people's memories of the landscape and the renewal and transformation of the symbols in the landscape (or in the cityscape). The places in our cases belong to the cities, not only in terms of cadastre, but also in terms of the social memory of a local community.

We are working with the landscape (and cityscape as a landscape in the city, because cities are located in the landscape) on two levels: on one level, as a geographical and ecological entity, a terrestrial mosaic of ecosystems, and, on another level, as a symbolic construction existing in the form of cultural and social conditions in a certain space and time. The key element for understanding landscape is a mutual interlinkage and unity between the place and community through culture. It is evident that changes are occurring within this unity and transform not only material forms, elements, patches, patterns, and mosaics of landscape structures but also the symbols themselves – the complexity of cultural construction in terms of what landscape was, what landscape is, and what we want landscape to be. Hills, trees, forests, and meadows – not to mention buildings, routes, fields, and ponds – are no longer considered as just hills or trees or forests, but they also bear symbolic meanings that, although not visible, are real in the cultural context. Our two cases are about discontinuation and continuation of the cityscape in both forms: symbolic and material.

From the early beginning of the first scientific definition of landscape by Alexander von Humboldt, we can see the connection between nature and culture in some specific forms. Around 1806, after his return from Chile, this great scientist, with his holistic way of thinking, expressed the following simple definition of landscape (in German): "Landschaft ist der Totalcharakter einer Erdgegend" ("Landscape is the total character of a region of the Earth" [Zonneveld 1995, 27]). He stressed not only geographic elements because, for von Humboldt, the total character of landscape contained all human and even aesthetical elements (von Humboldt 1995 [1814]). There are many definitions of landscape created by von Humboldt's transdisciplinary followers in the field of landscape ecology (Naveh and Lieberman 1994; Zonneveld 1995; Brandt 1999; Wiens and Moss 1999; Moss 2000; Naveh 2000; Opdam, Foppen, and Vos 2001; Tress et al. 2004). All these definitions define more precisely von Humboldt's "total character." The *European Landscape Convention* defines landscape as "an area, as perceived by people, whose character is the result of the action and interaction of natural and/or human factors" (Council of Europe 2000, 3). This definition is a synthesis of the previous ways of thinking, but it is not the final one.

Sociology also contributes to the definition of landscape, as explained in the treatise on knowledge and social constructivism by Peter Berger and Thomas Luckmann (1966). Even though they do not mention the words *landscape* or *cityscape* in their famous study, it is crucial for the symbolic construction of the landscape that their work unites Durkheim's and Weber's paradigms. Berger and Luckmann show the social construction of reality as an interaction among people in a certain time, and, more than that, they show how the construction of this reality becomes a real fact for the community through the process of internalization. As the authors stressed, "only with the appearance of a new generation can one properly speak of a social world" (55), and this, in turn, is important to understand a society's fear of the discontinuity of symbols.

Thomas Greider and Lorraine Garkovich (1994) go further in underlining the human impact on the landscape. For them, the landscape represents definitions of ourselves; groups of people socially construct landscapes, putting subjective symbols and cultural meanings into the environment. They write: "Cultural groups socially construct landscapes as reflections of themselves. . . . Thus, the natural environment and changes in it take on different meanings depending on the social and cultural symbols affiliated with it" (8). This interpretative framework of the landscape allows us to see a living landscape like a living symbolic system. On one hand, these authors are using the approach of C. O. Sauer (1925), stressing the activities of cultural groups in creating and maintaining the cultural landscape (although without quoting him as an influence). On the other hand, they are going beyond Sauer's concept of cultural landscape by advancing an understanding of the landscape within a process of symbolic interactionism, which is something quite far away from Sauer's attempt to interpret landscapes as an objective morphological entity. They note: "The framework of landscapes emphasises that what is important in any consideration of environmental change is the meaning of the change for those cultural groups that have incorporated that aspect of the physical environment into a definition of themselves" (Greider and Garkovich 1994, 19).

Landscape art (in short, land art) – and various forms of this notion (e.g. land art, Land-Art, Earth-art, or Earthworks) – is a kind of art using landscape, in its physical and ecological features, to create art in many forms, thereby forcing people to view the art in a context that is out of the museum and in the natural world. The landscape or cityscape is not just a place for the installation of land art, but it is a part of the artefacts, and it is material, a physical framework and art itself. Land art follows on from the prehistoric creation of art in/with landscapes, like messages carved into the stones or paintings in the caves. Modern land art appeared first in the United States in the 1960s, when artists began creating land works on a large landscape scale (Michael Heizer and his *Double Negative* in Nevada, 1969, is a good example). Today, modern land art expands all over the world. In many cases, the label *land art* is a confusing name for those engaged in gardening or landscaping or for landscape architecture companies involved in business activities.

It is common to use local natural material for creating new structures and symbols for land art. Sometimes, visitors are invited to contribute to the art in some way. Despite its large scale, land art is quite ephemeral in some forms, changing in the course of time and becoming, again, a part of the changing landscape. Land art artefacts have their own evolution in the landscape. Some of them persist for only a few years or months, or even for just a few days or hours, and this is the reason why documentation of land art is a part of land art creation.

Land art is often considered to be a part of site-specific art, or public art – that is, artwork created to exist and to have a connection with a certain place. It is connected with environmental art and landscaping and with architecture. Site-specific art also has its ephemeral forms – for example, site-specific dance (Kwon 2002). There are many artists involved in site-specific art, and some of them are considered to be founders of modern land art – for example, Henry Moore, Robert Smithson, Andy Goldsworthy, Christo and Jeanne-Claude, Michael Heizer, Walter de Maria, Richard Long, and many others.

Characteristics of land art

Six general characteristics tend to define land art:

Land art is a kind of protest. It is a protest again the total commodification of art: against the elite space of galleries for elite people, against the dictate of art dealers, and against consumerism. Artists are looking for a new common space for art and society.

Land art is a kind of respect for nature. The new space requires new art – this includes new (and old traditional) materials, concepts, and rethinking of the relationships between humanity and nature. Artists must respect the landscape; the shape of the wall created by Andy Goldsworthy at Storm King Art Center in 2000 can be interpreted in this way. Artists dedicate their art to the influence of natural processes on human life.

Some land art becomes a part of nature. A world-renowned example is *Spiral Jetty*, created by Robert Irving Smithson in 1970. Often lying under the water due to the fluctuating lake level of the Great Salt Lake (in Utah, United States), which is full of salt and animated by temporary sun and wind, the work reveals the interplay between human artefacts and nature with the changing seasons and natural powers. Land art becomes a part of the landscape in terms of natural ecological processes, as mentioned above, but also becomes a part of the cultural landscape too. This means that the works of land art become landmarks and symbols, for both individuals as well as collectives, and sometimes they even become internationally known artefacts.

Some land art installations work with the assistance of temporary natural powers. Walter de Maria and his installation of four hundred stainless steel poles works with lightning (*The Lightning Field*, 1974–1977, near Quemado, New Mexico, United States). Christo and Jeanne-Claude, from Bulgaria and France, with their *Valley Curtain* (1970–1972), encouraged people to imagine the landscape beyond the curtain as being mysterious and valuable. Their other installations are represented by the wrapping of natural or cultural patterns around various natural and human-made features and landmarks (for example, *Wrapped Coast* in Sydney, Australia, 1969; and *Project for Wrapping the Reichstag* in Berlin, 1995).

Land art is a kind of social network. The building of large conceptual land art requires the help of the community. This was the case for such projects as Robert Smithson's *Spiral Jetty* and Richard Long's *Karoo Crossing* in South Africa (2004).

Land art is a natural tool for human creativity on the landscape scale. From this point of view, it is more than settlements, architecture, gardening, landscape planning, or commercial landscaping. The intention of land art is to create symbolic forms, symbolic footprints, and structures using the landscape as a tool for this conceptualization.

Land, people, and art

Land, people, and art (LPA) is seen as a continuation of land art. It is a free, spontaneous creation of symbols of renewal using artefacts both to create the concept and to reinvigorate the memory of the landscape and the community. All the characteristics of land art mentioned above are also found in LPA, but with a stronger accent on social networking. This is a crucial point in the LPA process.

We witnessed the LPA process from its early beginning in the Czech Republic – that is, since 2012. Despite different locations and different local communities, both cases of LPA had a common trajectory. First, there was an attempt to erase the symbols representing the community before the communist era: to erase it (from real existence) in a tangible form as well as from the national consciousness. The easiest way to destroy the symbols of past history is to destroy them physically. Human history is full of these experiences. The same things also occurred in our case studies, not recently, but over a course of time together with the erasing of cultural continuity. For example, Königsmühle, as a village, disappeared in 1946 after the displacement of the Sudeten German inhabitants. The hill Zátiší ("still life") was abandoned in 1948 as a forest park and lost its functions as a meeting point because the citizens of the small town of Vodňany had no interest or motivation to maintain this place, full of memory.

Our cases

We have investigated two cases: one at Vodňany and Zátiší, and the second at Königsmühle (see Figure 15.1). In both cases, our research methodology included participant observation, literature comparison, and photo documentation, as well as thirty semi-structured interviews with participants in each project. After the installation of the land art, we visited the places in every season of the year to determine the condition of the artefacts and landscape and the number of visitors. These visits took place during weekends.

Vodňany and Zátiší

Vodňany (Czech for "water town") is a town in the Strakonice District of the South Bohemia Region. The town is located twenty-nine kilometers northwest of České Budějovice on the river Blanice. Its area is 36.35 km^2, its altitude is 398 meters, and it has a population of around 7,000 inhabitants (6,976 inhabitants in 2012). The former royal town takes its name after the word for "water" (in Czech, *voda*), which surrounds the town from all sides. In history, its "water magic" was mentioned in Medieval Latin texts as *Aquilea Bohemorum* (Czech Venice).

The first settlement appeared at the turn of the twelfth and thirteenth centuries, alongside the business route from Prachatice to Prague and also at the gold-bearing sites on the Blanice river banks. The Czech settlement was gradually transformed into the market town of Vodňany. The town is mentioned as a royal

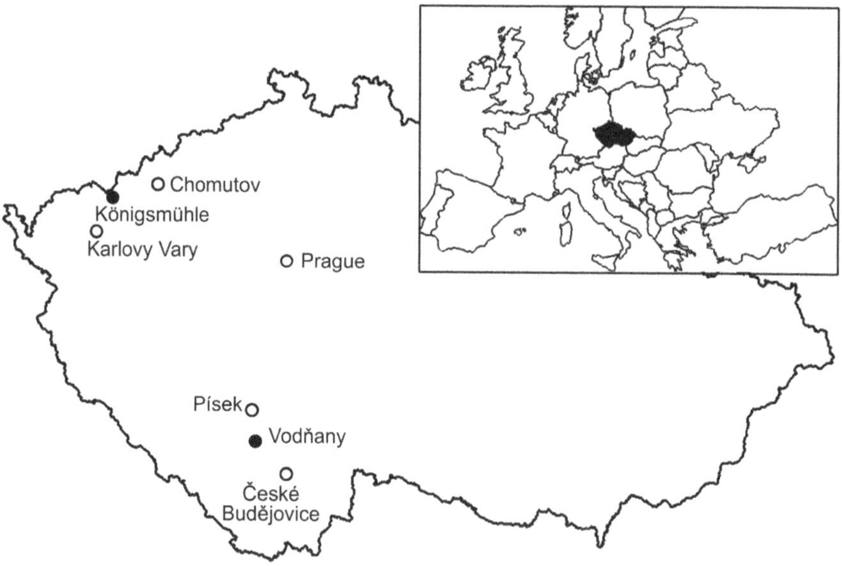

Figure 15.1 Map showing Vodňany and Königsmühle among other key places
Source: Map drawn by Josef Maxa, reproduced with permission.

town in relation to privileges given by King Jan Lucemburský (John the Blind) in 1336. Vodňany is well known as one of the traditional centres of the South Bohemian fisheries. In the year 1400, Vodňany together with the towns of Písek and České Budějovice were the pillars of royal authority in South Bohemia. Twenty years later, the Hussite leader, Jan Žižka, conquered the town, and, after that, the town remained for some time in the possession of the Hussites. After the Thirty Years' War, the town belonged to the noble family of Schwarzenberg.

For centuries, the city coffers were filled with profits not only from the old trade routes and payments of levying duties but also from the considerable income from the mining of precious metals in a place called Vodňanské Svobodné Hory. Remembrance of this mining activity remained a symbol of the town throughout its history from 1547 when it was promoted to a royal mining town. However, fish farming proved to be a more profitable activity. The municipality had already begun fishpond building in the second half of the fifteenth century, and the fishponds became a characteristic part of the landscape. The centuries-old tradition of fish farming is much more than just a past. The experience and knowledge of local fishermen are transferred to young generations at the Fishermen Secondary High School in Vodňany, which was founded in 1920. In 1953, the Research Institute of Fish Culture and Hydrobiology was established there, and its system of parcel experimental ponds belongs to one of the largest in Europe.

At the end of the fourteenth century, the hill Zátiší (Czech for "still life") was transformed from pastures into vineyards according to the privileges bestowed by King Vaclav IV. Later, at the end of the seventeenth century, the cherry orchard was transformed again into arable land, and at the beginning of the twentieth century a forest park featuring a promenade with sitting places and summer restaurants was built upon the hill. The decline of Zátiší dates from 1948, after the beginning of communism, to its final fading away by the end of the twentieth century, when the main roads from the towns of Písek and Prague were built, avoiding the historical road that connected the hill Zátiší with the town of Vodňany.

Königsmühle

Königsmühle (the original German name; in Czech, Královský mlýn, meaning "King's mill") was under the authority of the municipality Háj in the district of Chomutov. In 1860, fifty-three people lived there in seven buildings; five houses were residential, and two houses were, most probably, mills. The grain, which was transported from Bohemia to Saxony, was milled there. Now it is a natural protected area of 18.6 hectares at an altitude of 914–930 meters. A pathway to the small village, which comes from the direction of Háj, was an approximately 1.5 kilometer-long gorge. For centuries, an old trade route ran through it along which the grain, ore, timber, and other heavy loads from Bohemia to Saxony were carried using heavy horse-drawn carts. In consequence, the path was pressed deeper and deeper into the soil and now, in several places, reaches a depth of up to three meters. When the inhabitants of Königsmühle were expelled in the autumn of 1946, the remaining people from the Háj municipality used their empty houses as a source of building materials, and eventually the site, due to the gradual expatriation of the residents, was erased from the map.

The former settlement is a unique place, where the mutual penetration of landscape, history, and natural heritage can be seen. Königsmühle is composed of the residues of six buildings scattered about a shallow valley. On the right-hand side, one can find common building materials and the remains of farm buildings and equipment. To the left of the village, there are the remains of two mills, and, under the water tank of the first, it is still possible to see the remains of walls and a cellar. Even the water channel, or flume, which led the water into a small tank from which the water then turned a wheel before flowing back into the stream, is still visible in the surrounding terrain. In the case of Königsmühle, a set of significant historical buildings, open and passable terrain, and an aesthetic experience of the surrounding landscape are combined together in one place. Thus, it is not just about the wild natural valley, but it is also a preserved example of an extinct settlement in a mountain valley (*Land and Art Meeting* 2012).

The LPA process

LPA is a social movement; it is a social process of identification and renewal of the landscape or cityscape landmarks in terms of their symbolic forms. In this chapter, we analyse it as a chronological process and describe its main steps or milestones.

Step 1: identify the existence of a cultural imprint – that is, some marks in a local community's memories and some evidence in the form of a continuation of symbolic functions of the landscape. This initial step is the foundation of the LPA movement. However, these memories are rather fragmented: in pictures, postcards, narratives, and so forth, rather than being complete historical evidence. Nevertheless, former landscapes with their own particular landmarks exist despite the fact that their structural physical existences may have been destroyed. Of course, this cultural imprint does attract some people. In our case studies, it was narratives, memories, letters, and postcards as well as landscape paintings that attracted the relevant communities to think about the formal, social form of landscape renewal (reimagining the landscape).

Step 2: nongovernmental organizations (NGOs) *intervene*. NGOs and their active members seem to be the critical factor for building the network, with the help of many other local regional and international supporters. However, there has to be a concerted drive from informal leaders in the beginning to get things moving, and they often start from the local community because they are personally connected with a place and they resonate with the symbolic form of the landscape.

In the Vodňany-Zátiší case, the first step was made by the NGO Vodňany žijou (Czech for "Vodňany Alive") in 2010, followed by a workshop on architecture, "re:vodňany," in 2012, which was supported as well by the municipality. A similar situation was found in the place of Königsmühle. The initiative was opened by the NGO Anticomplex in 1998, followed by DoKrajin (Czech for "Into Landscape") in 2013. These NGOs were the opening actors for the whole process of renewal of the landscape, and their initiatives were focused on two or three public actions. It is not a necessity that these people are permanently working on the renewed places, and, in the case of Königsmühle, it would be impossible.

Step 3: network with volunteers. Symbolic landscapes are coming into existence by sharing narratives and ideas in many discussions, which are organized near or directly in the places that could be renewed. In Vodňany-Zátiší, a group of around one hundred volunteers, including mothers with their children, prepared a one-hectare piece of ground for the installation of land art objects, clearing it of natural seeding, building a path, and moving the grass (see Figure 15.2). It took about one week, so the individual volunteers changed during this time. The installation of the object itself was also possible thanks to the volunteers. People from all over the country were involved, using Facebook and other social networks to spread information and ask for help, but a predominant number of helpers came from the South Bohemia region, where Vodňany-Zátiší is located. The total cost of the whole project, including a daily allowance for volunteers and timber for the sculptors, was 172,000 CZK (approximately 6,500 euros), which was paid in the form of sponsorship by the municipality of Vodňany.

In Königsmühle, there was a similar situation with respect to the group of volunteers. A group of around one hundred people, mostly students of the Faculty

Figure 15.2 Vodňany-Zátiší. A. The ground is cleaned and ready for visitor access (work of volunteers). B. The installation *Fuller's Cupola* (named after the artist). C. The installation *View Point for Two*. D. View of the Zátiší hill through *Fuller's Cupola*

Source: A visit by the authors to Vodňany in 2012. Authors' own photos.

of Art at Charles University, Prague, cleaned the ground for the installation – cleaning an area of about 0.5 hectares, using the social network of the project leader, who was an alumnus of the faculty. Contrary to Vodňany-Zátiší, there were only a few local people as volunteers because of the abandoned Sudeten land and, therefore, a low population density in this area. As in the Vodňany-Zátiší case, there was an overwhelming number of young people (up to the age of thirty-five) involved. The budget was also very low – counting all the work and travelling, it was similar to that of Vodňany-Zátiší. In both cases, there were international sponsors: citizens from Vodňany and some German families who lived in this valley and surrounding villages until 1945.

Step 4: "Everybody is an artist." The LPA movement is characterized by the participation of volunteers in nearly every activity. All preparatory work for the land art objects' installation was completed thanks to the work of the volunteers. All objects in Vodňany-Zátiší – that is, *Fuller's Cupola* (named after the artist), *View Point for Two*, *Slow Worm*, *Fata Morgana Arbour*, *Apollo*, and *Ark* – were the projects of professional artists or students of art and architecture. What is appealing is the unifying idea: an environment-friendly, modern installation transforming a forgotten hill and its secondary forest back to a forest park again, and thereby renewing memories of the place.

There were twenty contributors connected with the land art installation in Königsmühle. In 2012, there were a lot of temporary objects, like a stone bridge crossing the stream, a view of the current village ruins through a painted village scene on a glass display board, a young portrait of the last surviving woman living in the village, and so forth (see Figure 15.3). Not all were material artefacts. "Do nepaměti" (Czech for "Into No-Memory") is the title of a site-specific theatre performance in which there were poetry and literature readings in the open landscape. Contributors were students or young artists, a few of them being professionals. The unifying idea was, for a period of one week, to bring alive the village of Königsmühle and in this process to renew the memory and narratives about the ruined buildings, thanks to the very dramatic life story of one displaced woman. Part of the project was to construct a real, although temporary, building that would serve as a meeting place (pub) in the ruins of the former mill.

Step 5: return to landscapes their symbols. This is possible by sharing experiences and symbols at a given place. The first step is information. NGOs use social networks – in the case of Vodňany-Zátiší, the municipality helped with the information flow, and regional newspapers were also used; in the case of Königsmühle, regional newspapers and a documentary on state television were used. Both cases used websites to make a rich documentation of the reconstruction of the places in the process of LPA. The increasing number of visitors is quantitative proof of the returning landscape. In Vodňany-Zátiší, the number of visitors grew constantly during a two-year period. In Königsmühle, there were waves of visitors attracted by workshops and performances during the one year of our case study visits; however, after these initiatives ceased, visits are made now only by a few individuals or small groups of tourists.

Figure 15.3 Königsmühle. A. Königsmühle today. B. Königsmühle during symbolic reconstruction – the pub. C. The decorated pub. D. Königsmühle during symbolic reconstruction – the village

Source: Königsmühle visit by the authors 2012. Authors' own photos.

In conclusion: is LPA more intervention or integration of landscape, people, and art?

To answer this question, the most important factor is the motivation. Motivation is explicitly expressed in the case of Vodňany-Zátiší by the keywords and phrases mentioned, like "renew forgotten days of the former glory of the forest park, connect the local community with the students and visitors of the town, stress the baroque axes of the cityscape and renew the nice perspective, renew the city promenade on the bank of the pond, viewing the town in relationship with its environment, rediscovery of the significance of the place, etc." (Gavriněvová and Machková 2013, 7). As Pavel Janšta has said, " 'Re' is the answer on the question what to do with Zátiší" (in Gavriněvová and Machková 2013, 13).

According to the project leader, Petr Mikšíček, the philosophy of land and art gatherings is to restore the memory of the place:

> We want to turn people's attention to regard Königsmühle as a cultural meeting place, to make an effort to put Königsmühle on the map again, to mark it on the tourist path and to erect an information panel about the history of the place. Also, to refresh the memory of the historical evolution of

> this landscape, respect to nature and today's citizens of Königsmühle, i.e., animals and plants and trees.
>
> (*Sdružení DoKrajin* 2014)

The aim of DoKrajin, the NGO that is dealing with Königsmühle, is "to rediscover the home in the places that were affected by the alienation, socio-cultural discontinuity, exchange of population, population displacement, land-use change, wild development or decline of regions" (*Sdružení DoKrajin* 2014). Leaders of the Königsmühle project use the term *documentality* to mean documentation by individuals of the significance of the places (*Sdružení DoKrajin* 2014).

As a means of contrast, it is interesting to compare this situation with that of *Tilted Arc*, a site-specific sculpture that was originally commissioned by the United States General Services Administration's Arts-in-Architecture programme for the Foley Federal Plaza in New York City. *Tilted Arc*, a model representing art-in-the-public-interest, was criticized for being "absolutely inappropriate to the site because of the top-down decision-making process, dictated by small review panels of art experts and bureaucrats, which did not involve the members of the local community" (Kwon 2002, 26).

It seems, for LPA, that a bottom-up approach is one of the main characteristics. The role of experts and professional artists is secondary; their role is to fulfill the idea of the renewal of the physical and symbolic characteristics of the landscape. There is another comparative situation: while *Tilted Arc* works with the concept of a public space that needs to be symbolized by new artefacts and meanings, renewal in the LPA process works with places full of meanings, which need to be rethought. In the case of public art paradigms by Kwon (2002), there are artefacts (sculptures) in the beginning, in their real existence, generating new symbols and new cultural constructions of the cityscape. In the case of LPA, there are intangible symbols of landscape in the cultural construction and primary meanings of local community, renewed by new artefacts located in the landscape. We can see two forms of discontinuity: tangible structural and intangible symbolic. The LPA movement's goals are to renew both of them, with new tools of land art and new artefacts and symbols, in respect to the ecological and historical environments. No institutional support from the EU or government is needed; it is more about NGOs, natural leaders of student communities, and volunteers and local communities.

LPA is using land art and other forms of installation in open public spaces, but there is a crucial difference: the LPA process is growing as a bottom-up movement – from leaders to community – from the beginning without support or intervention from government institutions. There are differences in our cases in the fine details of the motivation of people in LPA, but the main purpose remains: renewing memory of landscape and community.

LPA is the process of externalization of initiatives of endogenous sources on the individual level as well as in exogenous communities. The main purpose is to renew the social identity of the place and the local community in terms of social constructivism – the process of internalization of the landscape into an implicit, valuable, and symbolic entity.

The LPA movement is also about a reaction to the crisis of confidence in politics and its state institutions (Kunštát 2013). It seems to be natural to look for identification with something in the past, with the natural and cultural patterns. LPA offers new continuity, new identity, and new rethinking of individual and group intrinsic values through sharing the process of creating LPA.

An emerging problem seems to be the future of these installations. On one hand, the artists know their installations are temporary; on the other hand, there is a relatively large effort to keep this landscape in cultural construction by the local community. In the case of Vodňany-Zátiší, it has become a part of town planning because it is clear the forest park needs some ongoing maintenance. In Königsmühle, there is no permanent maintenance.

The main problem, according to the leaders of the LPA, is the acceptance of the local community. LPA is about renewal of the social function of art – improved cohesive forces, continuity, and social identity. To make this happen, LPA has used some modifications of land art, public art, and site-specific art. LPA is characterized by the prefix *re-*: revitalization, restoration, renewal, reanimation, and rethinking continuity of the landscape and community. Rethinking our cityscape means maintaining the integrity of the physical landscape and its symbolic cultural constructions. It also means increasing cultural sustainability and the resilience of a local community. LPA seems to be a relatively new and emerging process for how to reach cultural sustainability spontaneously, in a bottom-up fashion, using modern land art tools and their modifications.

Acknowledgements

The authors thank MSMT COST CZ support for LD141118 "Land and art cultural sustainability," COST Action IS1007, and Josef Maxa for the map and active investigation of sources.

References

Berger, P. L., and Luckmann, T. (1966) *The Social Construction of Reality: A Treatise in the Sociology of Knowledge*. Garden City, NY: Anchor Books.

Brandt, J. (1999) Geography as "landscape ecology." *Danish Journal of Geography* (Special Issue no. 1), 21–32.

Council of Europe (2000) *The European Landscape Convention*. Strasbourg: Council of Europe.

Gavrinĕvová, K., and Machková, M. (eds.) (2013) *Re:vodňany 2012 V Zátiší iniciace proměny zanedbané stráně nad městem – katalog architektonicko-sochařské dílny/ re:vodňany 2012 V Zátiší. Start of transformation neglected hillside above the town – catalogue of architectural and sculptural workshop* (in Czech with English summary). Vodňany: MAKE* detail.

Greider, T., and Garkovich, L. (1994) Landscapes: The social construction of nature and the environment. *Rural Sociology*, 59(1), 1–24.

Königsmühle land and art setkání/ Königsmühle (Land and Art Meeting) (2012) Available (in Czech) at: www.konigsmuhle.cz/aktualne-2012/ [Accessed 3 February 2014].

Kunštát, D. (2013) Důvěra ústavním institucím v prosinci 2013 [Confidence in constitutional institutions in December 2013]. *Naše společnost* [*Our Society*] (in Czech), 2–9 December 2013. Prague: CVVM SOÚ AV ČR, v.v.i.

Kwon, M. (2002) *One Place After Another: Site-Specific Art and Locational Identity*. Cambridge, MA; London: The MIT Press.

Moss, M. R. (2000) Interdisciplinarity, landscape ecology and the 'transformation of agricultural landscapes.' *Landscape Ecology*, 15, 303–311.

Naveh, Z. (2000) What is holistic landscape ecology? A conceptual introduction. *Landscape and Urban Planning*, 50(1–3), 7–26.

Naveh, Z., and Lieberman, A. S. (1994) *Landscape Ecology: Theory and Application* (2nd ed.). New York: Springer.

Opdam, P., Foppen, R., and Vos, C. (2001) Bridging the gap between ecology and spatial planning in landscape ecology. *Landscape Ecology*, 16(8), 767–779.

Sauer, C. O. (1925) The morphology of landscape. *University of California Publications in Geography*, 2(2), 19–53.

Sdružení DoKrajin/Association Dokrajin (in Czech) (2014) Available at: www.dokrajin.cz/clanek/17-poslani/sess_94c483ad04be5d2cc6cbad7f6bf1dbc0/ [Accessed 3 February 2014].

Tress, G., Tress, B., Harms, B., Smeets, P., and van der Valk, A. (eds.) (2004) *Planning Metropolitan Landscapes: Concepts, Demands, Approaches*. Wageningen: Delta Series 4.

von Humboldt, A. (1995) [1814] *Personal Narrative of a Journey to the Equinoctial Regions of the New Continent*. London: Penguin.

Wiens, J. A., and Moss, M. R. (1999) *Issues in Landscape Ecology: International Association for Landscape Ecology Fifth World Congress*. Snowmass Village, CO: International Association for Landscape Ecology.

Zonneveld, I. S. (1995) *Land Ecology: An Introduction to Landscape Ecology as a Base for Land Evaluation, Land Management and Conservation*. Amsterdam: SPB Academic Publishing.

16 Strumica, the city I want to live in!

Loreta Georgievska-Jakovleva and Mišel Pavlovski

Introduction: culture and local sustainability

At the start of the twenty-first century, attention to the role of culture in social, political, and economic development became more prominent, especially from the perspective of ensuring the sustainability of these areas of human activity. While public awareness about the possibility to redefine culture from a position as "consumer of a society's resources" into one of an "instigator of social inclusion and economic growth" has increased significantly, culture is much more than these instrumental roles. Importantly, culture is also "an expression of identity, power, and elaboration on relations that allow us to reach understanding with others" (Rogač Mijatović 2011, 331–340). In this chapter, from the perspective of contemporary Macedonian development processes, we investigate both how culture participates in and contributes to processes of sustainable development and how culture itself can be sustainable.

While sustainable development does not have a single, unified meaning, the definition put forward by the World Commission on Environment and Development (1987) serves as a general touchstone: "development that meets the needs of the present without compromising the ability of future generations to meet their own needs" (41). The future-oriented concept links the protection of the needs of future generations to the *adoption of appropriate decisions* today that aim to be valid for posterity. Such decisions are influenced by beliefs and values in the current cultural context, which, in turn, are informed and influenced by the past, intertwined in our collective and individual identities and cultures. As we move the emphasis from the decision itself to what has, or should have, decisive influence on the taking and adoption of a decision, we highlight the relation between culture and sustainability. Our decisions depend equally on what we were, what we are, and what we want to be. Our awareness of who we are allows us to understand the way in which our worldviews, experiences, and decisions differ from those of others. The term *other*, in our case, does not relate to racial, ethnic, or gender difference; it is tied to political and generational difference. This conceptual line positions place-embedded culture as a key dimension of the wider context for thinking about and making decisions in regard to our collective (more sustainable) development, and it also foregrounds the importance of simultaneously considering the past, the present, and the future.

Thus, this chapter encompasses two different perspectives: the role of culture in sustainable development (i.e. culture as the fourth pillar of sustainable development) and the sustainability of cultural practices (i.e. the sustainability of culture itself). We believe that the two perspectives cannot exist in isolation from one another, and the analysis of those questions can best take place through analyses of their mutual interactions (see Anheier and Hoelscher, in this volume). We contend that a well-designed cultural programme can develop local knowledge, talents, and ideas that contribute to higher quality of everyday life for residents and, ultimately, to the city's ongoing sustainability. This chapter aims to explore possible approaches to integrate culture in local sustainable development through an analysis of cultural practices that can embody and propel strategies for ensuring social inclusion and sustainable economic growth in small and medium-sized towns and cities like Strumica.

Cultural identity contexts

At a national level, the independence of Macedonia and the transition from a single-party to a multiparty system introduced many changes to the ideological matrices in the country. In particular, this situation brought an increasing focus on the stories "forgotten" during the socialist system – stories that raise the question of the right of the Macedonian people to culturally identify with the cultural heritage created on its soil in the age of antiquity. This directly confronted the existing position that insisted on the Slavic origins of Macedonians. Many believed that the "appropriation" of the antiquity heritage would mean for Macedonia a direct effort to dissolve the Macedonian identity. Therefore, an absurd situation was created in which the population was divided between those "for" and those "against" the so-called antiquisation, a matter directly related to the Greek veto on Macedonia's membership in NATO and the EU but which also had a direct influence on the quality of everyday life of citizens.

This division then entered the political discourse: SDSM, the party that held the power in the years after Macedonia gained independence, accused the current ruling party (VMRO-DPMNE) that the "antiquisation" of Macedonia undermines the very essence of the Macedonian people, who are identified with their Slavic heritage. In turn, VMRO-DPMNE accused SDSM of betraying national interests and working in favour of foreign interests. This situation creates tensions, which present serious obstacles for the development of the country – especially in cities like Strumica, where the local power is held by SDSM, while VMRO-DPMNE is in power in the national government. Locally, these tensions grow into severe divisions, even outright hostility, between members and supporters of the two major political parties.

In these disputes, nobody takes into account that identity is a fluid and dynamic concept, as reflected in Castells's (1997) definition of identity as "the process of construction of meaning on the basis of a cultural attribute, or a related set of cultural attributes, that is given priority over other sources of meaning" (6). Identity contains two (possibly conflicting) aspects: the old and the new identifiers constructed and suggested by current events. In the Macedonian case, the situation is doubly complicated: it is a conflict between the identifiers of the past, which are intertwined in current events and pretend to define the future, too. Should we accept that cultural identity is not a mere compound of ideas, beliefs, traditions, language, and customs accumulated over time but that it is also a cultural choice responding to external provocations or crisis situations in a given time frame, it is clear that the resulting crisis of identity matrixes (caused by internal and external factors) needs to be managed.

The cultural identity and cultural programme of the city of Strumica

The city of Strumica (population thirty-five thousand) is located in the southeast region of Macedonia, a relatively affluent agricultural region. It is an almost homogeneous ethnic community where the Roma and Turkish minorities enjoy, at least legally, all guaranteed rights and freedoms. Although employment and social assistance statistics may tend to indicate otherwise, class differences are not dramatic, and political and generational divisions are far greater challenges for the city.

During the 1990s, Strumica extended its interests from policies that base development exclusively on exploitation of natural resources to policies that pay attention to the ideas, skills, knowledge, and creativity of people as cultural capital that could bring economic and social development. Over time, Strumica has emerged as a centre for cultural life and vitality, securing a specific image related to its cultural identity. In April 2013, a survey (N = 200)[1] conducted with the citizens of Strumica and in other parts of the Republic of Macedonia found that the city of Strumica is widely recognized as a city with a specific cultural atmosphere (92 percent respondents) and a distinctive cultural identity.[2] The reasons given for the city's appeal ranged from "the local climate," "the proximity of the Mediterranean Sea," and "the possibility for better earnings due to the fertile land in the area" to "because there are things to do in Strumica." A significant number of respondents mentioned the city's "vibrant cultural life" or "fine entertainment and cultural life."

The city's cultural atmosphere was described as dynamic and lively (e.g. it is "a city that lives day and night" and a "city that is always full of people") and welcoming (e.g. the "hospitality" and a "city with a specific atmosphere where everyone feels comfortable," which was cited by all ten of the foreign nationals surveyed). A variety of the city's cultural assets were highlighted (e.g. "the city with the best disco-club in Macedonia," "the city of the carnival," and "its historic monuments"). A number of mental images and emotions generally come to mind when Strumica is mentioned, and respondents identified several things that they saw as specific to Strumica: its atmosphere, carnival, festival, and churches and monasteries; the Kale (the fortress); and, above all, the citizens of Strumica – the local residents who fill the streets, cafes, mall, clubs, parties, festivals, art colony, and all other cultural events with life and hospitality. The cultural identifiers (and thus the cultural identity) of Strumica consist of both the people and the events. While the lively and hospitable spirit of the local inhabitants can be found in other places in Macedonia, what makes Strumica special and distinctive from other cities of Macedonia is the rich and very important *cultural programme* that plays a main role in the life of the city.

The cultural identity of Strumica and its cultural programme are the result of a stormy and dynamic past that provides the basis for Strumica to be seen and experienced as a complex multicultural centre with specific characteristics and a specific cultural atmosphere. Perhaps subconsciously – caused by the practical need to ensure unity and social inclusion of the citizens, regardless of their background, in order to overcome the difficult transition – the question of cultural identity has turned into *an action plan* created through negotiations between the local population and the local and central governments that incorporates the elements and experiences of the past in a manner that corresponds to contemporary needs (Khaniki 2001). With the efforts of its civil society, an implicit cultural plan for Strumica[3] is emerging, created through the cultural programme which not only secures social inclusion and economic growth in the city but also connects the local and national governments and relates local cultures to global processes. Thus, the city's cultural identity and cultural programme can be viewed as important factors that have enabled local residents to overcome crises created by the transition and by historical misunderstandings, stabilized the city's identity, garnered international recognition for the region, and contributed significantly to the area's economic development.

Every year, Strumica hosts a variety of events, including the Carnival of Strumica, dating back to pagan times, which is held forty days before Easter, on the Tuesday after Quinquagesima Sunday (the Sunday before Lent); the international short film festival AsterFest, which presents the best Macedonian short and documentary films and a rich international programme; the Risto Šiškov Chamber Theatre Festival; the International Art Colony, which features about twenty artists, representing different cultural environments and artistic styles, in each edition; and the Strumica OPEN Festival, an event that promotes both high and popular culture. In addition, Strumica hosts many *ad hoc* cultural activities throughout the year.

The cultural programme: a space for social cohesion

In Strumica, the cultural programme implements inclusive cultural policies. This term is used to denote dialogue between different groups determined by their age, and, in regions with limited experience with political pluralism, between members and supporters of different political parties, with the aim to overcome social and ideological divisions. As mentioned previously, political and generational divisions are significant challenges for the city.

Overcoming political divisions

The political divisions create tensions and place obstacles to the everyday functioning of Strumica. The two opposing political groups do not cooperate, even on issues that are of common interest. This strict division along party lines does not only impede the development in the city; it lowers the quality of everyday life for its citizens. When it comes to Strumica's cultural programme, however, these divisions are overcome, and both the local and the national government give it full backing. The local administration negotiates with the national government, which provides part of the funding for organized events, while the national government supports local programmes as a way to win over voters' sympathies. As a consequence, events are funded by both national and local governments.

This "mandatory cooperation" results from the fact that cultural events and manifestations engage with all citizens of Strumica on an emotional level, regardless of their political affiliations. The feeling of belonging to one and the same cultural tradition, and the importance placed on contemporary artistic and cultural creation, contribute to a situation in which divisions, prevalent in almost all other spheres, are overcome when it comes to cultural matters. From a political perspective, cultural manifestations are places for two groups that see each other as enemies to successfully meet. The events provide opportunities for successful negotiations to overcome the tensions. At the same time, events promote creation and strengthen the feeling of belonging and participation among all citizens of Strumica, with 275,100 participants reported in 2014.[4]

Local manifestations and events are thus a good example of practices that can ensure common *participation in culture* of different (political) groups existing within one local community. While the views of each political group remain different, the programmes are articulated in a dialogue that results in cooperation. For example, in the carnival, masks are used, among other things, to express political views and positions or political disagreement. This allows citizens to actively participate in political life through cultural practices. Such dialogue between different political parties is not seen in any other place. Achieving concrete, creative outcomes is even more rare.

A different type of "cooperation" dynamic occurs in regard to festivals such as the AsterFest film festival, the Risto Šiškov Chamber Theatre Festival, or the International Art Colony. These are understood as elite national art with universal, humane messages and thus cannot be labelled as "products" with a political identity, something that often happens in other areas. Elite culture is widely viewed as the best that the spirit of a nation can produce and as the true representative of the state and/or ethnicity.[5] As a result, there is a generally held position that it cannot be labelled as a "product" of a political identity, something that often happens in other areas. Accordingly, elite culture stands on the top of the hierarchy of values, and all contributions to its nourishing and creation meet the widest approval. Regardless of their other differences, the political parties want to present themselves as supporters of the elite culture, something that is best evident in Strumica, a municipality in which the power at the local level is held by the opposition party. The central cultural events organized by the municipality – recognized and perceived by the local population as forms of elite culture – are financed from both the national budget and the budget of the local government. That situation pushes the political entities to cooperate with each other. It also illustrates the indirect influence of the citizens on decision-making processes, as well as possible approaches to establish communication between opposing political groups in other sectors.

Based on interviews with civil society representatives and focus group respondents regarding Strumica's cultural programme, and analysis of this case in the context of Strumica's politically divided citizenry, we identified two notable dimensions. First, the engagement of citizens' emotions, which arise from a feeling of shared culture and unity, contributes to more successful, inclusive policies. Second, culture is a place in which, through creative approaches (involving dialogue and collaboration), citizens can express different political opinions that, in another place or occasion, could be seen as threatening and hostile.

Multigenerational sensitivity

The realization that cultural manifestations bring certain benefits to citizens has led to Strumica's cultural policies being open to many types of initiatives. Lessons learned during the organization of the Strumica Carnival, for instance, have led to the acceptance of different perspectives and approaches and have made public policies open to initiatives coming from different groups. The idea to internationally promote elite Macedonian culture coexists with the initiative of a youth organization from Strumica to organize an event dedicated to popular culture, Strumica OPEN Festival, which was launched in 2012.

These observations place emphasis not on common traditions but on the important role that cultural history plays in engaging the emotions of those people who have a stake in the production of inclusive cultural policies. As Shalini Venturelli (2003) writes, the most important issue for any culture is not the heritage of its past but the inventive and creative capacities of its present. She notes that the most important issues we face today are related to the opportunities available to most people in society to take part in the creation of new cultural forms. Therefore, the circumstances in the environment most conductive to originality and synthesis, and the width of social participation in the creation of new ideas, are the true test of the strength of culture and are the sole valid foundation for a public policy.

The OPEN Festival is held in mid-August, the time of the year when about thirty thousand people who work and live abroad tend to visit their hometown. In 2012, the festival had a total of 220,000 visitors, and the 2013 edition had about 250,000 visitors. For 2014, it is projected that about 300,000 visitors could see the events held during the eleven days of the festival. The subsidies for this event are well above those of all other events. In the first year (2012), the municipality subsidized the festival with 2,500,000 MKD, and the subsidies were increased to 2,850,000 MKD in 2013.

From the current vantage point, following two successful editions of the OPEN Festival and with 220,000 participants in 2012, the municipality's decision to subsidize the festival seems logical. However, the circumstances today are not the same when the youth initiative was launched, and the funding negotiations can be seen as a case study in its own right. In Macedonia, especially in the smaller towns, the general belief is that cultural subsidies should focus on elite culture, viewed as the best that is produced by human spirit. (The ideas of the Frankfurt School and right-wing criticism of popular culture remain alive in Macedonia.) Popular culture is commonly seen as something consumed by the common masses and, therefore, ineligible for subsidies. It is often labelled as kitsch, as detrimental to good taste, and with a value that can be measured only by material profits. In spite of the fact that Macedonia has stars in all areas of popular culture, it seems that they all belong to the private sphere of individual preference. This attitude tends to lead to the production of "true" art that is boring and has no audience, usually supported with state subsidies.

In this context, the festival proposal had to be "packaged" in the proper form, as something that would engage the elite culture. Envisioned as a popular festivity of the Dionysian type, OPEN Festival, in the proposal stage, was decorated with elements of elite culture: a film marathon, literary and poetry readings, art happenings, and so forth. Such a proposal demonstrated not only the ability of the youth to negotiate but also their creativity in the resolution of problems generally arising from a generation gap. The result incorporated a popular celebration of local cuisine and Macedonian beer, a festival that promotes and places art in a new frame, so that it is no longer seen as useless and distant from the common person. The skill that the young people (as a part of civil society) demonstrated in the negotiations with the municipal administration further strengthened the culture of dialogue. The production of unconventional art, innovative creative approaches, and unexpected venues (e.g. poetry readings held in a public park) opened new thematic areas and types of encounters. The experiences of the past three years indicate that the main strategy has been successful. An official of the local municipality responsible for the event describes the event as bringing together tradition and modern living, and different life philosophies and lifestyles – on one hand, that which has been around for centuries, primeval and traditional for Macedonia, and, on the other hand, contemporary and modern. In addition to local food and beverages, the Strumica OPEN Festival project offers an excellent opportunity to enjoy music, from folk and traditional Macedonian music, to contemporary pop, rock, hip-hop, and electronic music. The Municipality of Strumica is the main sponsor of the Strumica OPEN Festival and subsidizes primarily music, theatre, arts, film, and folklore events (interview with Mr. Toni Milushev, Head of the Department for Local Development, Municipality of Strumica).

This example illustrates that an important dimension of the sustainability of a culture is its ability to adapt to changing conditions and ability to embrace the hybridization of art forms. In order to be sustainable, culture needs new forms of expression using the matrices of popular culture that will evolve or replace what are seen as old, outdated, and lifeless forms.

The city's overall cultural programme allows Strumica to gain local experience in bringing together the local administration and the principle of participation. According to Pindado, Izguierdo, and Olivé (2002), participation can be grouped into two broad categories: participation intended to ensure legitimacy, and participation for purposes of transformation. The experiences of Strumica can be placed into the second category: where the aim is "to strengthen the project that enables the citizens to propose and negotiate changes and transformations . . . the process of participation can grow into a true educational process for all workers that take part in it, including, of course, the public administration" (Pindado, Izguierdo, and Olivé 2002, 18).

Culture as an instrument of economic growth and brand for the city

Strumica's cultural programme is also a catalyst for the economic growth of the city, based on cultural resources and creative energies. The case of the cultural programme of Strumica makes it evident that culture and other creative fields are under the influence of social, political, and economic processes, while the economic processes are largely developed under the influence of cultural and creative forces. As an input, creativity is viewed as a driving and renewable energy that is not depleted by intensive use but, rather, that is increased through use. If the modern era expelled miracles from this world, the connection between creativity and economy in the twenty-first century restored them. Creativity is similar to the magic bag of gold from the famous tale – regardless of how much you spend, it remains full.

Many authors suggest that along with processes of transition from an industrial to a postindustrial society, traditional production factors are changing too. The production structure is being replaced with creative resources, and the future for economic growth is in the creative sector. Creativity not only contributes to economic value but also stimulates the formation of new ideas (and technologies) and processes of transformation and change (Isar 2013, 3). The quantity and the sustainability of cultural and creative energy depend on the ability to stimulate that energy to create a creative society that can harness creativity as a catalyst of economic value. As Florida (2002) argues, the promotion of creativity requires a strong creative class based on technology, tolerance, and talent – and the social and economic attractiveness of the location. The creative capital of a locale comprises the creative people that work there, a location characterized by an openness to new ideas, tolerance, and diversity. We commonly perceive such locations as *cultural brands*.

Strumica's cultural programme is also tied to the process of branding the city, aiming "to establish a significant and differentiated presence in the market that attracts and retains loyal customers" ("Branding" 2012). A well-designed campaign to brand a nation and/or a city is used not only to create an internal identity but also to create images intended for the rest of the world, in order to make a recognizable name for the nation/city and bring about economic profits. The policies of branding are especially in vogue in post-socialist countries. Countries, provinces, and cities have adopted the idea that the process of branding is "vital for attracting allies, investments, and earnings in the form of tourist dollars" (Gardener and Standaert 2003). The recognition of multiple, intertwined histories and identities underlies the more advanced branding processes. As Milena Dragićević Šešić (2009) has written, "every city . . . should develop not one, but many mutually intertwined identities, with the aim to use the best elements of its history, geography and contemporary human resources . . . and profile itself, above all, in line with contemporary values and resources" (122–123).

A cultural programme is seen as one way to present a city as such a product, with the potential for both cultural recognition and economic development and growth. Thus, the process of branding becomes closely related to building so-called culture clusters (Mommaas 2004). While Strumica's cultural programme is not a classic "campaign" (like the promotional campaigns at the national level), the ultimate aim of the municipal policies and promotional actions is to create an image of Strumica as a city that places culture at the focus of its attention, as a city of many festivals and cultural events, with dedicated and constant audiences – a coherent concept that is expected to bring economic and commercial benefits. The cultural programme builds an image of a city in which the creative potential of its residents and visitors creates an atmosphere that is specific to Strumica. The cultural programme is supported by an ongoing process of creation of logos, slogans, and other forms of advertisement.[6] These efforts create an image of a city that invests in culture and where local residents as well as visitors and tourists can share the creative energy of the city at any time of the year. This very same creative energy that is the brand of Strumica attracts tourists and is expected to attract foreign investors. As a result, for the citizens of Strumica, the cultural programme is a "public good" that unifies municipal, regional, and national interests.

The cultural programme has also influenced the physical public space of the city. As architecture is one of the cultural markers of a city, the central government invested in the development of a modern "European-style" central square. On one hand, this indicates the desire and political will to bring Strumica closer to the image of European cities, and, on the other hand, the square is envisioned as a venue for mass cultural events and scenes. In addition, the municipality invested in an amphitheatre in the municipal park, to be used as a site for smaller events. The goal is to create an image that will place the country and the city among the centres of cultural tradition and creative energy.

The success of that strategy can be illustrated by data on the number of visitors to the cultural events organized in Strumica (see note 4), revenues earned in the city through the events, and contributions of the cultural activities to job creation. Although the economic benefits of Strumica's cultural programme have not been comprehensively measured, Strumica has more than half a million visitors per year, which constitutes a significant source of income. That income is reinvested in programmes for the sustainable development of the city, such as boosting employment in the cultural sector (the cultural programme of Strumica directly employs fifteen persons and also contributes to job creation more widely). Although the number of direct jobs is small, the cultural events create opportunities for the local population to earn income through a variety of creative practices, particularly through the production of crafts. Local artists also benefit from the international dimension of cultural events, which allows them to connect to the international art scene, gain wider recognition, and secure better placement for their works.

Conclusion

If we reflect back on the research survey's finding that Strumica is a city where people want to live, we see that its vibrant cultural atmosphere and events are commonly identified as attractive assets and distinctive markers of the city. Underlying these positive perceptions, Strumica, perhaps unconsciously, eagerly accepts and embraces the concept of creativity. On the basis of developing people's skills, knowledge, ideas, and creativity as resources, and a willingness to invest in and promote culture as the "signature" brand and driving force of the city, Strumica has developed a reputation for its creativity and vibrancy. Within this, culture is treated not as an elite category for a narrow circle of cultural professionals and connoisseurs but as a dynamic category with huge potential for positive social change and for changes in societal relations. Cultural practices in Strumica show that, in addition to the known and established cultural professions (e.g. writers, curators, painters, etc.), new and equally important professions are emerging, such as craftspeople, caterers, managers, and so forth. A public awareness was raised that culture is not merely a "consumer of budget funds," a position that prevails in Macedonia, but a generator of social cohesion and stimulator of economic growth. In Strumica, a significant portion of the local population not only earns a living from culture, but cultural professions are self-sustainable thanks to the many visitors to the cultural events of the city.

In these mutually interdependent situations we see the relations between culture and sustainable development. Culture is no longer a mere consumer of funds but is now a factor of social cohesion and economic growth that, in turn, enhances the sustainable development of the culture itself. Based on the examples examined in this chapter, it can be concluded that decisions relating to cultural events, when taken on the basis of broad consensus, can influence both the sustainable development of a locale and the sustainability of the culture itself. The cultural policies of Strumica that put an emphasis on the realization of a cultural programme ensured the integration of culture in governance and laid the road to inclusive and sustainable development. In summary, Strumica's cultural programme illustrates a series of principles and practices that enable cultural development to align with and support the city's broader sustainable development:

- Build a general city culture that promotes social cohesion, including the construction and promotion of ethnic values, regardless of political and generational background, which, for the residents of Strumica, are the main points of separation and exclusion. This foundation capitalizes on the potential of the arts (especially the Strumica Carnival) to promote social cohesion.

- Processes of negotiation and dialogue can be facilitated through cultural/creative approaches. Political opponents – aiming to construct a picture for themselves as supporters of elite culture, viewed as representative of the best of a nation – accept joint participation in cultural events.
- Shared emotional attachment to place and to fellow residents can be mobilized through cultural activities, which can provide an inclusive approach to shared local cultural identification and contribute to social cohesion.
- The dialogue between generations within the city's cultural programme can stimulate new, hybrid art forms. Embracing new forms of expression that use and may evolve or replace older, less vibrant cultural forms can contribute to the sustainability of local cultures through adapted forms.
- The hybridization of art forms can increase the popularity of the art. Also, by using the methods of popular culture, elite culture may gain larger audiences and thus help ensure its sustainability.
- The culture sector can contribute to local economic growth and poverty reduction in manners that can nourish traditional cultural resources and contribute to new avenues for jobs and revenues. Economic development led by culture can preserve traditional crafts, protecting local cultural resources, which are often fragile and unique, nonrenewable capital threatened by extinction.
- Long-term investment in people's skills, knowledge, creativity, and ideas and investment in a variety of cultural activities can build new economic pathways for local development. Through culture, Strumica is growing into a brand name that attracts tourists and visitors, making the cultural programme a powerful subsector of the local economy, creating new jobs (mostly indirectly), and stimulating local development.

Notes

1 The questionnaire was sent by mail to 250 individuals, and two hundred were returned: ninety-five residents of Strumica, ninety-five citizens of other cities and towns in Macedonia, and ten foreign nationals residing in Macedonia.
2 When asked, "If you could choose, which city you would like to live in and why?" eighty-three of the ninety-five residents of Strumica and twenty-three of the ninety-five residents of other municipalities said they would choose Strumica.
3 Strumica does not have a formal cultural programme. Culture is planned and programmed through the municipal budget and through relations between the municipal administration and civil society actors. In this way, an implicit plan is emerging. It is subject to change, and it is dynamic to respond to the expressed demands and needs of the local community.
4 Responding to an inquiry from the authors in 2014, the competent bodies of the Municipality of Strumica reported the following visitor/participant figures for its cultural events: OPEN Festival, 220,000 visitors/participants; Strumica Carnival, 50,000; Risto Šiškov Chamber Theatre Festival, 4,000; and the AsterFest film festival, 1,100. These numbers refer to the average number of visitors annually for the period 2008 to 2014, a period during which the number of visitors/participants were estimated and assessed after each individual event was completed.
5 This is well illustrated by the political programmes of the major Macedonian parties, in which cultural programmes mostly refer to the so-called elite culture. This is even more

evident in the list of programmes financed by the Ministry of Culture, which almost without exception belong to the area of elite culture.

6 The unofficial slogan of Strumica's cultural programme, "Home of Culture," was borrowed from a blog site and is a poor English translation of the name given to municipal facilities of the socialist era.

References

"Branding" (2012). Businessdictionary.com. Available at: www.businessdictionary.com/definition/branding.html [Accessed 29 October 2014].

Castells, M. (1997) *The Power of Identity: The Information Age: Economy, Society and Culture, Volume II*. Oxford: Blackwell.

Dragićević Šešić, M. (2009). Kultura u funkciji razvoja grada (Kulturni kapital i integrativna kulturna politika). *Kultura*, no. 122–123, 20–40.

Florida, R. (2002). *The Rise of the Creative Class*. New York: Basic Books.

Gardener, S., and Standaert, M. (2003, March) Estonia and Belarus: Branding the Old Bloc. *Brandchannel*. Available at: www.brandchannel.com/features_effect.asp?pf_id=146 [Accessed 29 October 2014].

Isar, Y. R. (2013) *Creative Economy Report 2013: Widening Local Development Pathways*. New York; Paris: United Nations Development Programme and UNESCO. Available at: www.unesco.org/culture/pdf/creative-economy-report-2013.pdf [Accessed 22 November 2014].

Khaniki, H. (2001) Role of media in identity formation. *Hamshahri Daily*, no. 2606, 12–13.

Mommaas, H. (2004) Cultural clusters and the post-industrial city: Towards the remapping of urban cultural policy. *Urban Studies*, 41(3), 507–532.

Pindado, F., Izguierdo, Q. R., and Olivé, J. M. (2002) *Eines per a la participacio ciudadana: Bases, métodes i técniques*. Papers de participacio ciutadana, no. 6. Barcelona: Diputacio de Barcelona.

Rogač Mijatović, Lj. (2011) Kulturne aktivnosti Srbije u Evropi i svetu: 2000–2010. *Kultura* (Beograd), no. 130, 331–340.

Szondi, G. (2007) The role and challenges of country branding in transition countries: The Central and Eastern European experience. *Place Branding and Public Diplomacy*, 3(1), 8–20.

Venturelli, S. (2003) *From the Information Economy to the Creative Economy: Moving Culture to the Center of International Public Policy*. Cultural Comments Series, Center for Arts and Culture, Washington, DC. Available at: http://goo.gl/Y5dgHZ [Accessed 29 October 2014].

World Commission on Environment and Development (1987) *Report of the World Commission on Environment and Development: Our Common Future*. Available at: www.un-documents.net/our-common-future.pdf [Accessed 29 October 2014].

17 Artistic/design practices, art education, and sustainable development

A question of design or of mutual transition?

Nelly van der Geest

Introduction: contributing to a sustainable city

In October 2012, the City of Utrecht, which aims to become climate neutral by 2030, signed covenants on carbon emissions with clusters of firms and organizations with shared interests, including the three institutes of higher education in town. HKU, the University of the Arts Utrecht, is now challenged to define its position in relation to these sustainable city ambitions. Beyond changes to its operations to become a climate-neutral institution, the situation provides an important opportunity to rethink its curriculum and its role as a "sustainability educator." Unlike the two other educational partners, however, HKU does not have an educational track that is directly related to the ecological dimension of sustainable development, such as technics or life sciences.[1] How can HKU give "life" to the city's environmental sustainability ambition within the core of its own processes – namely, educating artists and designers? Towards this end, how can the interaction between sustainable development and culture (and, in particular, art) be better understood?

Utrecht is a middle-sized city in the heart of the Netherlands.[2] The city has three universities – the University of the Arts Utrecht (HKU), Utrecht University (UU), and Utrecht University of Applied Sciences (HU) – with a combination of about fifty-seven thousand students, of which about four thousand study at HKU, which is centred in the inner city (the other two are located on the outskirts of the city). As a student city, Utrecht is sensitive to fashionable urban living and lifestyles. Art students are often early adaptors or even precursors for new lifestyles and, through this, contribute to the changing identity of the city.

Creativity forms the core element of the diverse spectrum of professional education in art and design. It is a fluid concept, and Robert Sternberg's 1999 definition of *creativity* is a useful starting point: "producing work that is both novel (i.e. original, unexpected) and appropriate (i.e. useful, adaptive concerning task constraints)" (cited in Hoogeveen 2014, 126). This definition speaks of the combination of novelty and utility, and it seems to accentuate a problem-solving approach. Nirav Christophe (2014), inspired by Andreas Reckwitz (2013), argues that a diverse array of forms of creativity occurs simultaneously nowadays.[3] Both Christophe and Reckwitz perceive an impossible demand on today's citizens to be creative. At the same time, however, Christophe emphasizes the importance of a collective element of creativity for today's artists. He argues that the contribution of artists should not be limited to an aesthetic product and that the work of an artist is shown not only in products but also in the process, during which artists can reflect and comment on the systems they use. These options to broaden traditional thinking about creativity beyond originality and appropriateness appear to be highly relevant for acting within the context of sustainable development.

As a starting point for defining *sustainability*, I use the widely cited Brundtland definition for sustainable development, which aims to meet "the needs of the present without compromising the ability of future generations to meet their own needs" (WCED 1987, 1). What might be the role of artists and designers in constructing a sustainable city? What are possible contributions of artists and designers to the reshaping of sustainable city life in the twenty-first century? Two perspectives on contemporary cities offer insightful frames for considering these questions. First, Charles Landry (2014) describes a city as a complex organism in constant movement with changing perspectives, opinions, and priorities (see also Landry, this volume). Cities are not closed systems, and the experiences of twenty-first-century city life are invented and lived out by citizens in the context of an open and dynamic city. Landry proposes seven interlinked concepts that are reshaping twenty-first-century city life: shared commons, eco-consciousness, healthy urban planning, the idea of the intercultural city, the aesthetic imperative,[4] creative city making, and invigorated democracy. These dimensions propose core values for the future sustainable city. Concerning invigorated democracy in particular, we have to take into account that thinking about encounters in public space has shifted from the idea of a free and equal discussion between strangers towards that of a performance in which individuals stage their identity and create personalized lifestyles (Geest 2013). Artists and designers can play important "lifestyle modelling" roles in this regard.

Second, from a more "procedural" perspective, Sacha Kagan (2011) argues that sustainable development is not a centrally controlled process based on a few key players but a multiplayer model of change. In a multiplayer situation, each party strives for its own values, perspectives, and gain. Communication becomes central in such processes. Successful communication of viewpoints organizes communities and helps to drive the process toward the values of the player. Groups of citizens with a certain lifestyle – such as young artist-designers – can become such players in the city. Kagan focuses on the process of sustainable change. Systems, especially when kept implicit, largely determine the outcome of sustainable development. Interventions in these systems – clarifying the way a system is organized and empowering new communities to become significant players in it – might change the mechanisms that up until now have tended to lead to unsustainability.

While Landry proposes the values that support a sustainable city, Kagan advises us to take the systems into account and to open them up for new communities to enter. Together, these two perspectives highlight the creative, entrepreneurial, and collaborative skills that should be integrated in the education of artists and designers in the contemporary period, times which require heightened attention to an array of sustainability issues.

Designers' quest or mutual transition quest

How are artistic and design practices relating with sustainability? Considering the interaction between sustainable development and design/artistic practices, I observe (so far) two options for relating these fields: the *designers' quest* and the *mutual transition quest*. These two poles are helpful to assess practices of artist-designers as they tackle issues of sustainability. Of course, the division of artistic/design practice in the context of sustainability into two quests is a theoretical separation only, and I use these quests as cornerstones for discussion. A gliding scale between the two options is probably more reflective of the broad array of real-life practices in play today.

The designers' quest

Traditionally, designers and artists contribute to the communication of concepts. In the *designers' quest* scenario, design – and perhaps even fine art – helps sustainable development to express itself in various forms of communication. Sustainable development determines the content, and artistic practices are used as a means to communicate or shape this content. In other words, sustainable development makes use of artistic practices, and artistic practices serve the goals of clients with a sustainability agenda. The main focus of the art is the form of communication, and the main issue of sustainable development is the content. The fields of expertise are combined but do not influence each other. The premises of each field are not changed by the connection.

In the designers' quest, knowledge of sustainable development is not fundamental for artistic and design practice itself. It poses preconditions to the process, but it does not influence the artistic process itself. Sustainable development is just one potential content. Clients determine the relevance of the issue. It is the individual choice of a designer to take sustainable development into account. The surplus value that art and design contribute is just the packaging, which can be applied to any social or economic question. The role of artists in society is not fundamentally changed.

Along this trajectory, attention to sustainability issues is generated when clients request (ecological) sustainability. Introducing clients into the curriculum is more often the case in applied art – especially in design for communication, architecture, and product design – than in fine art, performing art, and music, which are thus less likely to be touched by questions of (ecological) sustainability within the current curricula. From this perspective, introducing sustainable development into a curriculum is externally motivated, driven either by clients, the peer pressure of fellow universities, or sensitivity to trends. In this picture, the dedication of HKU to the topic of sustainability is volatile.

An example: Imagining Tomorrow

Imagining Tomorrow (http://imaginingtomorrow.hku.nl) is the annual interdisciplinary collaboration in which two hundred students of HKU take part. The event has a pressure-cooker approach to developing new concepts, meeting the needs of firms and organizations within the Utrecht creative industry sector. Within two weeks, interdisciplinary student-teams have to develop an answer to a question a firm or organization poses in relation to the main theme: "green economy." There are ten clients involved. Each client is assigned three teams of students to answer their "green economy" question, and the teams compete in the event. Each client nominates one of the teams for the long list. A professional jury chooses the overall winner.

It is Friday, early morning, towards the end of the event, and all thirty student groups have to present their work to the firms. The huge hall is divided in tables where each group exposes its solutions. There is a broad audience of the participating students and clients but also interested colleagues, both students and lecturers. The presentation event has a festival atmosphere. I see models of buildings, visuals of possible communication, designs of props in public space, and promotion movies of concepts. I see clients deeply involved with their three student teams, interviewing them about the concepts they propose. Later on, I meet at the coffee machine with lecturers and find clients intensively ranking the three solutions they were offered: "The strength of this proposal is the way they meet our demand to help us to form a community, their solution to connection." "But this one is so funny and at the same time simple and clear, this definitely will work," disagrees the colleague.

After an open market for visiting all the proposals, the audience is gathered for the central presentation of the nominations. Each client pitches its problem during a live interview and nominates one of the three teams that conceptually addressed the question. A video is shown in which the students introduce their concept. In the interviews, the clients explain the criteria used for nomination. It is striking that all firms claim it was a close call. They are very content with the proposals. They mention criteria such as how well the proposal met the needs of the firm; how well it took the unconscious culture of the firm or its clients into account (a bonus was acquired if the proposal also connected to other targets of the organization); the feasibility of the concept; and the proposal's contribution to the sustainability of the firm. This last criterion was variously interpreted: some firms approached sustainability from a planetary health point of view (environmental), others identified sustainability as contribution to the continuity of the firm (economic), and others took (social) connectedness with communities of customers as their central focus point. Beyond these "typical" criteria and the usual debates on the content of the sustainability concept, there is one criterion that is most decisive, shared by all commissioners: the creativity or "fun factor" of the concept. There is also a tendency to nominate those teams that invite the firm to go beyond its usual assumptions and perceived limits.

In 2014, the winning concept of Imagining Tomorrow was *The Twelfth Man Illuminated* (see http://imaginingtomorrow.hku.nl/?p=1266). Football Club Utrecht was the client. The stadium they own has an enormous energy bill, and they wondered how they could make use of the natural energy of their audiences to decrease their energy needs. The solution the students offered was to install technical sensors, such as piëzo crystals that take the kinetic and audio energy of an audience and transfer it into electricity.[5] These crystals were put under the seats. The electricity gained is used to illuminate "the twelfth man," the section where the fans are seated. A built-in aspect of the concept is competition, an element that fits the rituals of football fans. The transfer of kinetic energy into electricity is already technically developed at this time,[6] but the application to transfer audio energy to electricity does not yet exist to our knowledge.

The strength of the winning team was to combine an existing technical solution (even to propose to expand it) with a communication strategy, tailor made for the target client. A second strength was the involvement of the audience. The art students showed a keen eye for the audience perspective, and they searched for interaction. By the strength of the illumination generated, the audience will know they make more noise and movements than the section of the opponent. Students took into account that the eco-consciousness of football fans would not be rational (Hoekstra, Verheggen, and Hannink 2014), so the interaction of the audience was focused not on views of sustainable urban living but on adding something playful as a bonus for their enthusiasm. The "extra value" the proposal brought to the sustainable development perspective was attention to fun as part of the communication strategy on eco-consciousness.

Imagining Tomorrow contributes to raising an implicit eco-consciousness in its audience. From this perspective, Landry's "aesthetic imperative" is a key value. Playfulness or "fun" is embodied in the design and changes the connotation of the concept of *sustainability*. However, the project does not focus on reflecting on systems provoking *un*sustainability or on empowering new players to enter societal processes and systems. The skills to play with sustainability – and directly intervene in larger systems – are not (yet) passed over to the audience. Imagining Tomorrow 2014 is a good example of the designers' quest approach to linking artistic/design creativity with sustainability issues.

The mutual transition quest

Considering the change the industrial revolution brought to the role of artist and the underlying concept of creativity – namely, from skilled craftspeople into freethinkers who mirror society – it is likely that sustainability will likewise become a changing force for art and design. Thus, the *mutual transition quest* scenario is based on the conviction that both sustainable development and artistic practices are changed by their interaction. The mutual influence causes, in both fields of expertise, new processes, goals, and (perhaps) values. In the mutual transition quest, the inner ethical and professional processes that shape art education are redefined, and artistic practices are changed. The interaction changes the role of artists and might stimulate innovation in artistic practices and within art education. Vice versa, artists and their work influence the frames of thinking about sustainability and could bring innovation to debates on sustainable urban living. From this perspective, sustainable development is not just a content but a dynamic force behind changes in artistic practices and in art curricula.

An example: De Ceuvel, Amsterdam North

De Ceuvel is a polluted brownfield, a former wharf in Amsterdam North. Due to the economic crisis, the area could not be regenerated into a housing area, and the municipality launched a contest for the most innovative use of the brownfield for the upcoming ten years. Part of the contest was to design an environmental remediation and development plan in which the local government would not invest.[7] The contest was won by the architect Sascha Glasl and his company, Space&matter, an Amsterdam-based office for architecture, urban planning, and concept design. Glasl notes:

> As architecture and planning touch on all social and cultural levels of society, we think our reach and responsibility goes beyond form and function. Rather than just being reactive, we like to initiate projects and proactively involve ourselves in matters that we consider relevant. Setting up joint ventures around our own interests and fascinations has proven to be an effective way to realize our ambitions.
>
> (http://nl.linkedin.com/in/saschaglasl [Accessed 22 August 2014])

Together with Metabolic – a firm for applied research in the circular (closed-loop) economy focusing on food, water, energy, and sanitation – and landscape designers Delva, Space&matter developed a plan to create a "prohibited garden" with soil-cleaning plants. In that garden, roads of bamboo or wood form the infrastructure and connect old houseboats. The houseboats where chosen because of the heritage of the wharf but also because they are mobile and could move on to new experiments after ten years. The boats were obtained at no charge: individuals donated the boats to avoid the recycling tax (about €4000) if they were disposed. All boats were outfitted with a circular sanitation system. One of the boats became a lab in which human excrement and urine would be used for gas for the café. Over time, "the former industrial plot will be turned into [one of] the most unique and sustainable urban developments in Europe. The site, which is now heavily polluted, will feature imaginatively retrofitted houseboats placed around a winding bamboo walkway and surrounded by an undulating landscape of soil-cleaning plants" (http://deceuvel.nl [Accessed 22 August 2014]).

The place is intended to become a hatchery/incubator for artists, so an association was formed of the creative and social enterprises that will use the area. The association that manages De Ceuvel is the overall steward of the territory but also invests in linking the different firms to strengthen their clout and to attract assignments and clients. Members offer their free time and energy to recycle the old houseboats into offices, ateliers, or workshops and to run the association. In this case, a network of players works together. The players that are active in De Ceuvel comprise a mix of young artists and small creative enterprises with a few middle-sized, more established companies such as Delva and Space&matter. The task Space&matter took on was not just architectural but also to develop and remediate the area: they widened their role. Due to the approach taken by Space&matter, new players combined their interests and became stronger in realizing new settings.

The plan is complex and has different layers. There is an ecological dimension: the reuse of wasteland, the forbidden garden to clean the ground, and the circular system for sanitation. There is an aesthetic aspect: the creation of a built environment – the architectural side. And there are new forms of collaboration in play, embodied in the association to realize the ownership of the area and to interweave the firms. Here, social, artistic, economic, and environmental aspects interact. A lot of the processes are experimental. Will the bio-cleaning method work? Will the association be able to involve the owners in a continuous and democratic way? Will the association lead to greater economic resilience for the individual firms? Will meeting other creative and social companies lead to new collaborations, products, or processes?

Last but not least, the tenants at De Ceuvel create and live out an attractive lifestyle: eco-conscious urban living. In June 2014, the café was opened and immediately attracted lots of attention. The café is run by a collective that aims to make all processes in the café circular, serve only biological food, have a roof garden for growing vegetables, prepare their own juices, create the gas for cooking through the sanitation lab, and build tables and chairs of leftover materials. The place is becoming a popular spot to lunch and lounge. In a very short while, Café De Ceuvel gained 9,455 likes on Facebook (at 25 August 2014). In October 2014, De Ceuvel won the Frame Public Award of the Dutch Design Week (http://delva.la/nieuws/de-ceuvel-wint-frame-public-award [Assessed 20 November 2014]).

Many of Landry's concepts reshaping twenty-first century (sustainable) city life are happening on this site: shared commons, eco-consciousness, healthy urban planning, creative city making, an aesthetic imperative, and invigorated democracy. The integral approach of the site also redefines the conception of cleaning polluted urban areas: it is no longer a prerequisite after which housing can be organized, but it becomes part of urban living. The process of cleaning is part of a dynamic urban living. In this case, I observe the influence of artistic processes on sustainability by directly designing and taking on the cleaning process and changes in the roles of artists through their involvement in sustainability thinking: the creative process becomes collaborative. The focus is both on living out the new values of sustainable life (Landry) and on changing the concepts of working and living in an urban environment (reflecting Kagan's system-oriented views). In these ways, the project opens up the urban lifestyle for new interpretations. This case shows tendencies towards the mutual transformation quest.

Some theorists see the core task of artists and designers in the twenty-first century as offering new values for life. Michiel Schwartz and Joost Elffers (2010) claim that *Sustainism* is the new modernism. As the concept of *modernism* changed the values to be lived and sought after in the twentieth century, the authors predict that sustainable questions will reshape the values to live with in the twenty-first century. In the *Sustainist Design Guide*, Schwarz and Krabbendam (2013) promote sharing, localism, connectedness, and proportionality as core values. These value proposals oppose the old value proposals of modernist city life. *Sharing* opposes having – that is, the need to possess as a token of success in life. *Localism* opposes the non-rooted globalism of modernist city living. *Connectedness* questions egocentric individualism, and *proportionality* proposes attention to appropriate scale: "small and slow" as an alternative to "bigger and faster."

In August 2014, I visited De Ceuvel and observed a tendency of the artists and designers to live out these new values and to offer a new urban lifestyle – and by that, change the role of artists and designers in the city. The entrepreneurs and artists I talked with confirmed that, to them, the experiment was very inspiring and challenging, and it changed their mindset about urban living.

Implications

To distinguish between the (traditional) *designers' quest* and the (emerging?) *mutual transition quest* scenarios, four items were considered:

1 Are the interventions that artists and designers propose connected to one of the interlinked concepts of Landry (2014): shared commons, eco-consciousness, healthy urban planning, the idea of the intercultural city, the aesthetic imperative, creative city making, and invigorated democracy?
2 Which kinds of players do the projects support? Do they intervene in the system and empower new communities (Kagan's focus)? Do they support the established players, or do they provide entrance to new communities and new views?
3 Do they bring new topics or perspectives to thinking about sustainable development?
4 Can we notice new practices in creative processes or in the way artists and designers perceive their role?

The main difference between the two approaches is that, in the first scenario, sustainable development is an optional and temporary issue for art and design practices, while in the second scenario, it will have irreversible influence on artistic practices. From a perspective of "autonomy," the distinction becomes more profound. In the designers' quest option, designers would support the communication of clients with a vested interest in a sustainability process, whereas in the mutual transition quest option, it is open as to whom the artist or designer supports and whether he or she gives impetus to new communities that want to become more prominent players in advancing sustainable development.

These reflections connect to a core element of the training of artists and designers – the centrality of creativity in artistic practices – and the way HKU introduces students to phases of creative processes. Karin Hoogeveen (2014) connects creative expression to three factors: high motivation, the capability to enjoy the process of creation for its own sake, and concentration. In training creative people to become either designers or artists, developing a personal drive that highly motivates them is crucial. A personal drive brings them the authenticity on which they can build an individual profile to anchor their "doing" and "making" activities. By knowing their drive, they can develop a public identity. This element connects to the dynamics of a sustainable urban development in which everybody can play their part in proposing priorities and connecting broader audiences to them.

An individual example: Coleoptera, by Aagje Hoekstra

HKU offers product design education as one of its educational tracks. Within product design teaching, although not required in the curriculum, we see a number of designers who conduct research on potential materials. For example, Aagje Hoekstra, a 2013 graduate of HKU in product design, was interested in material that, at first sight, seemed to be waste but from which she wanted to create beauty and utility (Hoekstra 2013). Insects – for eating as well as other applications – have always fascinated her. She discovered that meal beetles, which produce mealworms to feed cattle, died after three months of production and became waste. She liked their shields for their transparent beauty. She started to peel these and interested a chemistry student to help her to find out how to make material from the shields, which turned out to be made of chitin. In a chemistry lab, she extracted the chitin from the shields and made a bioplastic from it. With this insect-plastic, she made light objects and jewellery called Coleoptera, peeling every beetle by hand (see www.aagjehoekstra.nl/coleoptera.php [Accessed 22 August 2014]). She became publically known through the attention she received during Dutch Design Week, and she received unexpected feedback from physicists and chemists who explained how she could improve the process of obtaining the raw material (see unst.hku.nl/648?prev=1 [Accessed 22 August 2014]). She is now starting a firm to improve the raw material and to make unique products from the bioplastic. She also organizes events to give people an opportunity to touch the insects and invites the audience to peel the beetles.

Hoekstra is a designer who begins with a personal fascination but does not limit this fascination to just producing products. She uses the processes and the products as integral elements in her communications to the public. In this way, she widens the practice of designers.

In her work, she is not steered by an external client with a sustainable agenda. She gives form to eco-consciousness by her actions. Through her search for the beauty of nature, she contributes to people's views on beauty, which I connect to Landry's aesthetic imperative. Due to the attention to beauty in her work, it blurs the division between artistic practices focused on inner drives, concepts, and beauty, and design applications focused on originality and utility.[8]

With the Coleoptera project, I observe a moving away from the designers' quest in the direction of mutual transformation. I perceive a change in the role of artists. The division between applied and free art is blending, informed by what artists envision as a better place to live, starting from their own fascinations. In the creative process, beauty and utility are connected, and the artist seeks attention for beauty within sustainable development.

Creativity in sustainable cities: some observations

To understand what is happening in the encounter between sustainable development and artistic practices at the current time, I introduced three Dutch practices: Imagining Tomorrow, an interdisciplinary conceptual design competition; De Ceuvel, an area development experiment in Amsterdam run by artists and architects; and the Coleoptera project, by product designer Aagje Hoekstra. In each case, I considered whether there is interaction between artistic practices and sustainable development, aiming to distinguish any tendencies indicating an emerging mutual transformation.

The three cases show indications of a slight movement towards *mutual influence* between artistic practices and sustainability. The aesthetic imperative and increasing eco-consciousness occurred in all cases. Attention to the systems that produce (un)sustainability parallels attention to the creative processes through which sustainable products can be developed. In *The Twelfth Man Illuminated*, the main focus was on the communication product. In the Hoekstra case, there was attention to the serendipity in the process, which offered her a new product (bioplastic) and motivation to connect people to insects. In the De Ceuvel case, many of Landry's visions about the emerging twenty-first-century city are being converted into practice. From an artistic/design perspective, there appears to be a widening of professional practice by combining process and product as part of the practices. This could be an early indicator of mutual change.

Artists and designers seem to pay more attention to collaborative creativity and offer new models for eco-conscious urban living. At the same time, designers make use of individual creativity (as it is trained in fine art) to develop propositions for sustainable living. This influences the role of designers in sustainable contexts. In a design practice, a broader range of concepts around creativity and creative processes seem to support the designer. Art education could deepen this to prepare designers for sustainability. Implementing this in design curricula proposes a new mission for institutional players as HKU, a university of art.

The cases I examined for this chapter did not shed light on the relevance of sustainable development for fine art or performing practices (i.e. non-applied arts). In these areas, sustainable development might be content of an art work, but is it influencing artistic practice? The blurring of artistic/design approaches in Hoekstra's case may be at stake in design, but is it also at stake in fine art or performing art?

Tendencies towards a richer concept of creativity in the cases indicate possibilities for mutual impact between sustainable development and creativity. The cases suggest that for developing new views on urban living, linking one's individual, internal drive to collective purposes is needed. For insights into systems, a process orientation is helpful. However, more research is needed to legitimize the integration of sustainable development within art education through its influence on the core processes of art creation. Does the blurring of practices that is happening in design also occur in the fine and performing arts domains? How do artists and designers analyse the influences the sustainable agenda has on their creativity?

In closing, I propose eight elements of creativity that might be relevant for the new roles of designers (and maybe artists) who want to work in a sustainable urban context. Inspired by Csikszentmihalyi (1996), who proposes that it is a quality of the creative person that he/she copes with oppositions in a creative process and knows when and how to switch from one aspect to the other, these elements of creativity oppose each other. Thus, the eight potential core elements of artistic practices in the context of urban sustainability incorporate individual *and* collaborative creativity; product *and* process orientations; and serendipity *and* disciplined approaches and capacities in craft and in concept. The final duality is the capacity to switch between an internal, individual drive and an external, collective focus, which should be a core capacity of artists/designers working on issues of urban sustainability.

Art educators should expose artists and designers to all elements of this multifaceted creativity so that they simultaneously develop both individual (authentic) processes and collaborative creativity and learn to cope with both internal and external orientations at the same time. Developing forms of "creativities" enables artists and designers to enter into and address the unplanned aspects of sustainable cities and the symbolic geography (Reckwitz 2010) with which cities develop a distinctive identity. With insights into creativity as a value that fosters the unknown, the unplanned, the intuitive, the emotional, and the coincidence, and that recognizes the environment as driven by its own rules, designers and artists can influence the dynamics of cities toward constructing more aesthetic and liveable places.

Notes

1 HKU offers education in a wide range of artistic practices, from fine art to design, from theatre to games, from art education to autonomous practices. HKU distinguishes its contribution to arts education through a strong emphasis on innovation and cultural entrepreneurship, and it has a tradition in social issues related to entrepreneurship. The unifying elements of its different tracks are queries around developments in creativity. None of the educational tracks of HKU takes sustainable development as a core subject.

2 In 2014, Utrecht had 330,772 inhabitants (source: www.stadindex.nl/utrecht [Accessed 26 August 2014]).

3 Nirav Christophe (2014), referring to Andreas Reckwitz (2013), lists the following as core elements of creativity: an individual creation process; an aesthetic utopia; creativity as liberating and revolutionary force; creativity as connected to suffering; creativity as useless, the Homo Ludens; creativity as collaborative; creativity as a staging process of product and artists' (private) lives; and the reflection of others (in new media) on both product and artist.

4 The term *aesthetic imperative*, to my understanding, is a quality criterion of the environment: making an environment attractive for all senses, so living becomes agreeable in that surrounding.

5 Vlooswijk (2014) reports in *Volkskrant* that the University of London has found a system to transfer sound waves into energy through a piëzo (zinc oxide) material.
6 See, for example, the Sustainable Energy and Dance Floors at www.sustainabledanceclub.com/products (Accessed 26 February 2014).
7 After the Municipality of Amsterdam awarded the plan, it later provided some financial support.
8 Hoekstra is not the only designer/artist to seek to stretch the limits of design in the broader context of environmental consciousness. In the Netherlands, Daan Roosegaarde is also famous for this (see www.studioroosegaarde.net [Accessed 22 August 2014]). He proposed "smart highways" by making the signal lines on the road with paint that could hold the energy of the sun by day and light up by night, eliminating much of the need for other lighting.

References

Christophe, N. (2014) *Creatief Vermogen Utrecht, 21–5–2014* [video]. Available at: http://vimeo.com/99554800 [Accessed 25 August 2014].

Csikszentmihalyi, M. (1996). *Creativity: Flow and the Psychology of Discovery and Invention*. New York: Harper Perennial.

Geest, N. van der (2013) Who's voice? Interventions in public space. In N. Duxbury (ed.), *Animation of Public Space Through the Arts: Toward More Sustainable Communities* (pp. 97–114). Coimbra: Almedina.

Hoekstra, A. (2013) *vpro.nl DDW13 report: De insectenmeisjes, Aagje Hoekstra & Evelien Crooy* [video]. Available at: www.youtube.com/watch?v=BT7hrvWb4do [Accessed 26 February 2014].

Hoekstra, J., Verheggen, P., and Hannink, M. (2014) *5 tinten groen (5 Shades of Green)*. Amsterdam: Motivaction Research and Strategies. Available at: www.motivaction.nl [Accessed 26 February 2014].

Hoogeveen, K. (2014) Creatief zijn of creatief denken. In N. van der Geest and A. De Boer (eds.), *Creatief Partnerschap, evenwicht tussen creativiteit en samenwerking* (pp. 125–138). Amsterdam: International Film & Theatre Bookshop.

Kagan, S. (2011) *Art and Sustainability*. Bielefeld: Transcript Verlag.

Landry, C. (2014) Civic urbanity. Available at: http://charleslandry.com/themes/civic-urbanity [Accessed 26 February 2014].

Reckwitz, A. (2010) *Culture and the Making of Worlds* [video]. Available at: www.youtube.com/watch?v=PAHtTASvyx8 [Accessed 22 August 2014].

Reckwitz, A. (2013) Die Erfinding der Kreativät. *Kulturpolitische Mitteilungen*, no. 141. Available at: www.kupoge.de/kumi/pdf/kumi141/kumi141_23-34.pdf [Accessed 22 August 2014].

Schwarz, M., and Elffers, J. (2010) *Sustainism Is the New Modernism: A Cultural Manifesto for the Sustainist Era*. Berkeley, CA: d.a.p.

Schwarz, M., and Krabbendam, D. (2013) *Sustainist Design Guide*. Amsterdam: Bis Publishers. Available at: www.sustainabledanceclub.com/products [Accessed 26 February 2014].

Vlooswijk, E. (2014, August 23) De week van de wetenschap. *De Volkskrant* [Booklet], Sir Edmund, no. 21, p. 2.

World Commission on Environment and Development (WCED) (1987) *Our Common Future*. New York: United Nations. Available at: http://conspect.nl/pdf/Our_Common_Future-Brundtland_Report_1987.pdf [Accessed 26 August 2014].

Index